SPACE OBSERVATORIES

ASTROPHYSICS AND SPACE SCIENCE LIBRARY

A SERIES OF BOOKS ON THE RECENT DEVELOPMENTS
OF SPACE SCIENCE AND OF GENERAL GEOPHYSICS AND ASTROPHYSICS
PUBLISHED IN CONNECTION WITH THE JOURNAL
SPACE SCIENCE REVIEWS

VOLUME 21

JEAN-CLAUDE PECKER

Professor at the Collège de France

SPACE

OBSERVATORIES

SPRINGER-SCIENCE+BUSINESS MEDIA, B.V.

LES OBSERVATOIRES SPATIAUX

First published by Presses Universitaires de France, Paris
Translated from the French by Janet Rountree Lesh

Library of Congress Catalog Card Number 70–124847

SBN 90 277 0168 7

ISBN 978-94-010-3322-0   ISBN 978-94-010-3320-6 (eBook)
DOI 10.1007/978-94-010-3320-6

INTRODUCTION

Le ciel est, par-dessus le toit,	The sky is, up above the roof,
Si bleu, si calme!	So blue, so calm!
Un arbre, par-dessus le toit,	A tree there, up above the roof,
Berce sa palme.	Waves leaves of palm.
La cloche, dans le ciel qu'on voit,	A church bell, in the sky I see,
Doucement tinte.	Softly tolls.
Un oiseau, sur l'arbre qu'on voit,	A bird, upon the tree I see,
Chante sa plainte.	Sadly calls.

PAUL VERLAINE

Like Verlaine, we are in prison. The prison is our Earth, "which is so pretty"; our atmosphere and its clouds, its "marvellous clouds". (You would think that Verlaine, Prévert and Baudelaire had been comparing notes!)

The sky is up above the roof... A tree there, up above the roof... Stars in the sky, like birds... their rays, like bells (and here we are with Apollinaire!)

What we see opens the way to what we guess at; what we observe leads us towards the unobservable. A poem releases images, and the invisible grows big with reality.

Astronomers are a little like poets (indirectly from the Greek ποειω, *make*): they *make* the universe by interpreting messages, extrapolating spectra, and inventing 'models' of the cosmos or of stars – fictional constructions whose observable part constitutes only a small fraction of the whole, and which only the inductive logic of the theoretician allows us to consider as representing unique physical reality.

The extrapolation of the tree we see and the extrapolation of the stellar spectrum we observe are more reliable if the treetop is larger, the spectral range wider. And this is why astronomers have for decades considered their information to be insufficient, and have wanted to extend their observations to new regions. A hundred years ago, only visible radiation could be observed: violet, indigo, blue, green, yellow, orange, and red. Then it was the ultraviolet, the infrared and, more recently, the radio region. But these windows for astronomical observation, opened by ground-based laboratory techniques, are too narrow, too limited. And to open them a little wider, today's astronomers launch observatories in space – balloons, rockets, satellites, space probes (Figure 1) – an attempt to escape!

In this book, we shall try to summarize the why and the how. The problems posed by the observation of the new space vehicles themselves – considered as artificial celestial bodies, objects for study by ground-based telescopes – are treated in another book in this series.* Here we shall deal only with the use of telescopes installed on these spacecraft with the object of *observing* the very natural universe around us, but not of becoming part of it or even of perceptibly approaching it.

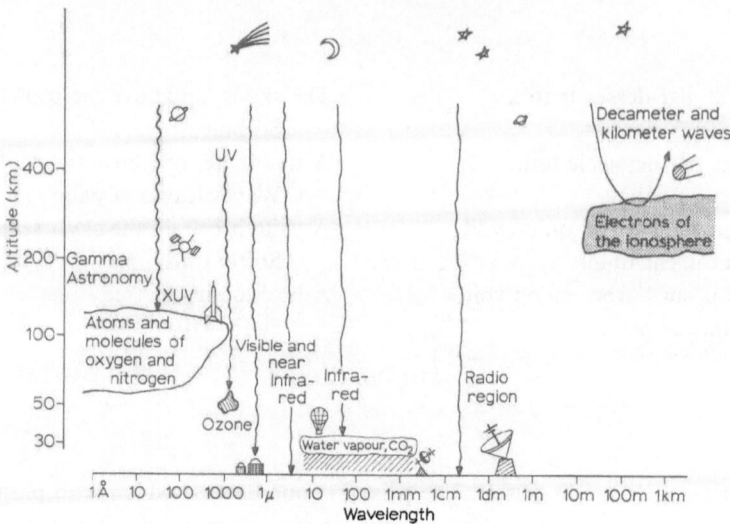

Fig. 1.   The windows for astronomical observation. – For a long time, the only usable window was that of the visible region. Progress in reception techniques for electromagnetic waves has opened the near-infrared window, and that of the short-wavelength radio region. But to open up the gamma-astronomy, 'XUV' and ultraviolet regions, rockets are needed to by-pass the absorption of oxygen, nitrogen and ozone; opening the infrared window requires balloons, at least, to rise above the opaque layers of water vapor and carbon dioxide; and decameter and kilometer waves are observable only from outside the ionosphere... Definitions of the wavelengths, spectral regions, etc. will be found in Chapter II, Section 1.

Why then put telescopes on rockets? Why put spectrographs on satellites?

Between the comfortable observatories of Haute-Provence or California and the thousand fires of the universe, the Earth's atmosphere scrambles communications. The air, opaque to so many kinds of radiation, seriously perturbs the few that it lets through. The task of the imprisoned astronomer is a subtle detective game – decoding the message, reading between the spectral lines, inventing the invisible.

How can the air (the popular expression, undiscerning as usual, is 'thin air'!) surrounding the Earth – that shallow layer of air only a few kilometers thick – influence and even paralyze astronomical observations by its presence alone, forcing the imprisoned astronomer to escape?

We shall try to answer this question, first by describing the different layers of the

* J.-C. Pecker, *Experimental Astronomy* (transl. by Robert S. Kandel), D. Reidel Publ. Co., Dordrecht, 1970.

atmosphere as we know them today, then by analyzing their properties with regard to the propagation of light (and, more generally, of all kinds of electromagnetic waves), and finally by showing how the atmosphere acts as a peculiar kind of screen to particles coming from space.

The extra-terrestrial laboratories have just begun to deliver something besides promises. In the second part of this book, we shall present some of the recent results, without going into the techniques by which they were obtained (that would take volumes of explanation!). We do this in the knowledge that at the very moment of publication these results will be quite out of date, and with no other motive than to suggest the future of this research, which promises to be exceptionally fruitful. The author will be pardoned if he puts special emphasis on the first part (the *raisons d'être* of space astronomy), which may have some permanent value; the reader should refer to the bibliography, intended as a permanent source of articles to appear as well as of published works, to obtain the latest results of the projects we have reported in progress.

Acknowledgements

The author wishes to thank Madame Déchard, who did part of the typing; Madame P. François and Monsieur P. Faucher, who completed with great care the difficult construction of the original Figures 3, 4, and 5 (Part I), and the calculations of Chapter III (Part II); and finally, Mademoiselle G. Drouin, who put the finishing touches on the manuscript with great competence and devotion.

I am deeply indebted to my translator, Dr. Janet Rountree Lesh, who has not only made a fine translation, but also improved the original on several points, pointed out a few mistakes, and made it when possible up-to-date.

TABLE OF CONTENTS

PART II / A PROVISIONAL AND PARTIAL INVENTORY
OF SOME OF THE INFORMATION TO BE ACQUIRED
BY SPACE RESEARCH

PART I

THE *RAISONS D'ÊTRE* OF SPACE ASTRONOMY

CHAPTER I

THE STRUCTURE OF THE ATMOSPHERIC LAYERS

1. The Development of Our Knowledge

Since the time of Pascal and Torricelli, who ventured (already knowing the weight of air) to measure the pressure of the atmosphere in centimeters of mercury and to ascertain its variation with altitude, the science we now call *aeronomy* has, indeed, made a certain amount of progress. Pascal's decisive experiment, at Puy de Dôme, took place on September 19, 1648. As a result of this experiment, he managed to calculate the total mass of the air surrounding the Earth. His treatise on the weight of the air mass concludes with these words:

Therefore the entire mass of the sphere of air around the world weighs 8 283 889 440 000 000 000 pounds ... that is, eight million million million, two hundred eighty-three thousand eight hundred eighty-nine million million, four hundred forty thousand million pounds.

This value (Pascal's pound being equal to 707 g) is almost as good as could be calculated today, namely 5.30×10^{21} g. But since air is far from having the homogeneity attributed to it in Pascal's time, this calculation would not be of much interest!

So the barometer and the balance were the first weapons of aeronomy. It would be unjust not to mention theory next. First Pascal and then Laplace applied the theory of hydrostatic equilibrium to the atmosphere. A 'Laplace atmosphere' is a model atmosphere of uniform temperature, in which the density decreases exponentially with increasing altitude h according to the formula

$$d = d_o e^{-h/H}$$

where H is the 'scale height', on the order of 8 km at ground level.

There followed, still at ground level, the chemical study of air. It is no doubt superfluous to dwell upon the importance of work by Priestley, Lavoisier, and Scheele on the properties and composition of air – at ground level. One-fifth oxygen, the rest nitrogen, and some 'impurities'.

In the nineteenth century, aeronauts had precise experimental objectives. Their balloons were always equipped with barometer-altimeters. But these considerations were generally secondary to the sporting aspects, and making measurements was the means rather than the goal of the flight. As early as 1804, Gay-Lussac and Biot directed expeditions for the French Academy of Sciences – Gay-Lussac himself ascended to an altitude of more than 7000 m. The variation of the water-vapor content with altitude, the distribution of the magnetic field... Measurements made at different altitudes, later complemented by unmanned sounding balloons and then by rockets and satellites, made it possible to determine the temperature distribution and to correct the Laplace atmosphere.

Any human message that travels through the air can be used to explore it, just like a balloon. The wireless – to give it the name it had at the time – immediately demonstrated the existence of the ionosphere (by *fading* and long-distance reception). The name of Sir Edward Appleton is connected with this discovery. How much progress has been made since then! Towards the middle of the 20th century, geophysicists were directing a major part of their attention to the ionized layers of the ionosphere, their connection with solar activity, and their extensions far from the Earth. Rockets have complemented the results of radio-wave soundings. Beyond the ionosphere, we now know of the Van Allen belts, the geocorona of atomic hydrogen, and the magnetosphere with its outlines deformed by the solar wind.

It is not our purpose to write a history of the science that has become aeronomy, nor to explain the origin of the ionized atmosphere or the details of sounding methods. But these few lines have shown the complexity of the atmospheric medium. For the astronomer, to be sure, the result of these investigations is not an isolated piece of information: is not the atmosphere, for him, to some extent part of the universe? The ultraviolet radiation of the Sun governs certain properties of the ionosphere, the X-radiation certain others... However, for the purposes of this book, the reader must excuse us for not beginning with a treatise on aeronomy, and for passing directly to the results of the analysis.

2. The Altitude Distribution of Density, Pressure, and Temperature

Figure 2 represents the 'mean' atmosphere adopted in 1961 by COSPAR, as regards temperature and density. The values indicated are only 'mean' values. The diurnal and nocturnal temperatures – and therefore the corresponding pressures – are very different from one another. The dashed curves give an idea of the amplitude of these variations.

Moreover, solar activity has a distinct influence on these properties, especially the density. The upper layers are obviously the first to be affected by variations in activity, since the ultraviolet and X-radiation from the Sun travel through these layers and are absorbed in them.

It is important to realize the difficulty of formulating a theory for such distributions, even in the 'model' atmosphere. Although we might assume that, on the whole, hydrostatic equilibrium is nearly satisfied (despite the winds, the convection, the turbulence...), a problem of almost inextricable complexity is presented by the absorption of solar radiation in the various layers, whose physical state governs the absorption but depends in its turn on the equilibrium to be reached. Thermodynamic equilibrium cannot be assumed and, moreover, the conditions are not stationary: what is absorbed during the day is radiated by night.

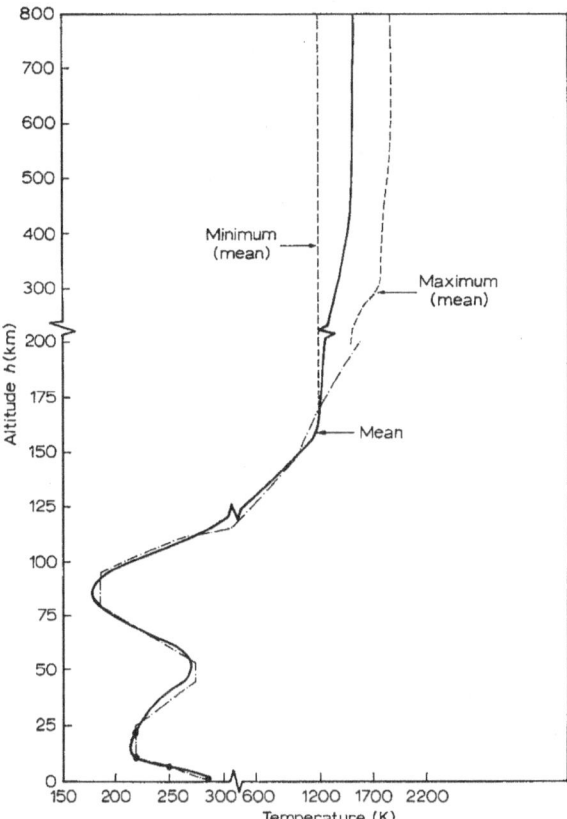

Fig. 2. The temperature in the reference model of the terrestrial atmosphere (after COSPAR, 1961). – *Solid line:* mean (at 45° latitude) derived from the data of the American scientists; *dotted and dashed:* from the data of the Soviet scientists; *dashed:* mean values of the extreme temperatures, according to various observers (the temperature distribution in the atmosphere varies with time, and from one geographic point to another).

3. The Distribution of the Principal Atmospheric Absorbents

At ground level, and without taking into account the considerable fluctuations in certain constituents (water vapor and carbon dioxide), the composition of the atmosphere is given in Table I.

Of course, the composition changes with altitude. In the first place, the formation of ozone is facilitated by ultraviolet radiation at altitudes from 15 to 30 km. Even more important, the pressure becomes low enough at very high altitudes for molecular nitrogen and oxygen to be dissociated. This phenomenon is obviously encouraged by the concomitant increase in temperature. Finally, in addition to dissociation there is ionization of the molecules and atoms; the number of atoms ionized remains small, but the ions and free electrons form the ionosphere. With its large opacity for radio

TABLE I

Composition of the atmosphere at ground level

Constituents	n (particles/cm^3)	$\log_{10} n$
N_2	2×10^{19}	19.3
O_2	5.4×10^{18}	18.73
H_2O	3×10^{17}	17.48
A	2.4×10^{17}	17.38
CO_2	8.5×10^{15}	15.93
Ne	4.7×10^{14}	14.67
He	1.35×10^{14}	14.13
H_2	1.285×10^{13}	13.11
N_2O	1.285×10^{13}	13.11
CH_4	2.5×10^{12}	12.4
O_3	4.725×10^{11}	11.675
O	1.05×10^4	4.025

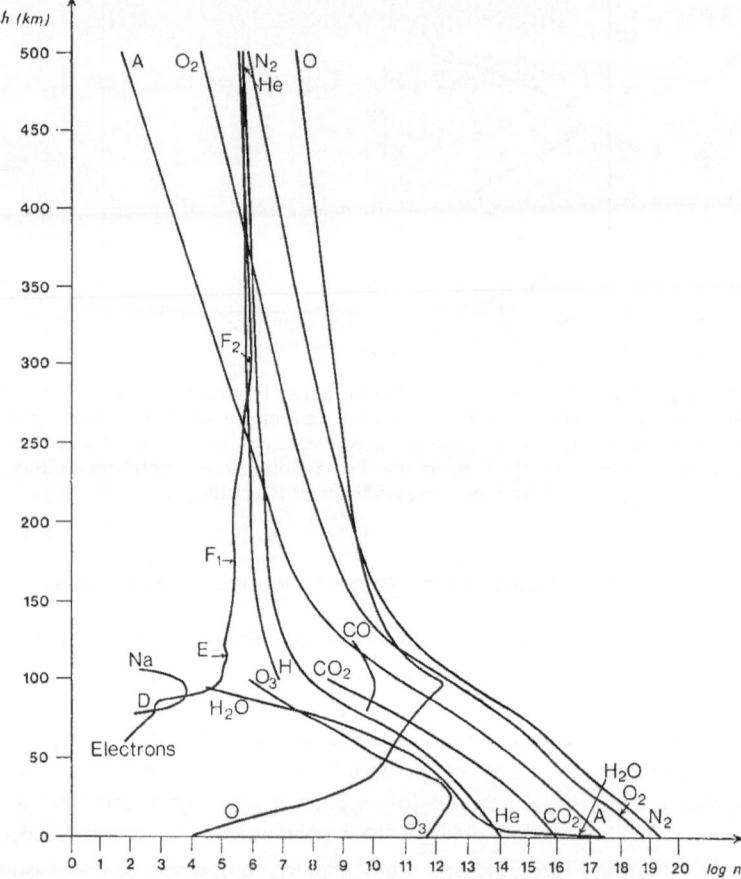

Fig. 3. Variation with altitude of the various constituents of the atmosphere. – *Abscissa:* the number of particles per cm^3 for different atomic or molecular gases.

waves and its high index refraction at long wavelengths, the ionosphere prevents radiation of wavelength greater than about 10 m from passing through the atmosphere.

Figure 3 shows the variation with altitude of the various constituents of the atmosphere.

In general, then, the principal perturbing effects of the atmosphere depend upon the type of radiation in question. Table II gives a general idea of these effects.

TABLE II

Wavelength range	Principal physical phenomena	Astronomical effect
$\lambda < 3000$ Å	Rayleigh scattering Atomic and molecular absorption	Optical thickness $\gg 1$: *Opacity*
3000 Å $< \lambda$ $<$ a few microns (visible) (near ultraviolet and near infrared)	Rayleigh scattering	Optical thickness generally < 1: *Blue of the sky* *Partial opacity* *Refraction* *Deterioration* *of images*
A few microns $< \lambda <$ mm	Molecular absorption	*Opacity*
mm $< \lambda <$ dam (radio astronomy)	(Rayleigh scattering) Electron scattering	*Diffusion* *Refraction* *Deterioration* *of images*
$\lambda >$ dam	Electron scattering	*Refraction* *Reflection*

Note that the properties of refraction and degradation of images are closely linked to the partial opacity in the two 'windows' (visible and radio) of the spectrum (see below, Chapter II, Section 3).

It is therefore with a study of opacity that we shall begin our demonstration of the inherent weakness of ground-based astronomical observations.

CHAPTER II

THE OPAQUE WALL OF THE ATMOSPHERE

We recalled just now that in the popular expression, air is an example of insubstantiality... thin air! And in fact, the terrestrial atmosphere is insubstantial and quite transparent to the solar radiation, the the light of the stars... But to be exact, the air is transparent only to radiation in the visible and neighboring regions (and also to certain radiation of long wavelength, between a millimeter and a decameter). We shall see that this restriction is quite sufficient! But first, let us see how we distinguish the different types of radiation, and what we mean by the 'spectrum' of a star or of the atmosphere.

1. The Spectrum

We know that the light emitted by celestial objects (as by any other light source) is not simple. It is composed of photons, with which are associated waves. The photons differ from one another in their energy, and the waves differ from one another in their wavelength – the wavelength λ of the wave being inversely proportional to the energy E of the associated photon.

Wavelengths are expressed in centimeters (in the c.g.s. system of units). We shall use this convention in all the algebraic expressions given in this book. But in practice other wavelength units are used, depending upon the application at hand: the angstrom unit ($1\ \text{Å} = 10^{-8}$ cm), the micron ($1\ \mu = 10^{-4}$ cm), the millimeter, and the meter.

Instead of wavelengths, we often use frequencies. The following formula (in which c is the velocity of the waves in question – that is, the velocity of light) enables us to transform from one to the other:

$$v = c/\lambda.$$

The unit of frequency is the hertz (or 'cycle per second'). Radio astronomers use frequencies more often than wavelengths, which are more familiar to astronomers working in the classical optical region.

The photon energy, $E = hv = hc/\lambda$, is also very frequently used in the gamma-ray region. For units, we use not only the erg but also the electron volt, the energy required to move an electron across a potential barrier of 1 volt.

Finally, the number of waves per unit length, $s = 1/\lambda$, expressed as a number of waves per centimeter (cm^{-1}) – is often used by molecular spectroscopists, who work mostly in the infrared. Atomic and molecular energy levels in spectroscopic tables are often given in wavenumbers (cm^{-1}).

Tables III and IIIa summarize the relations connecting these units, and the usual designations of the principal wavelength regions.

TABLE III

λ (cm)	λ (customary units)	$\nu = c/\lambda$ (customary units)	$s = 1/\lambda$ (cm^{-1})	$E = h\nu$ (eV)	Designation
10^{-10}			10^{10}	8.06 MeV ⎫	
10^{-9}			10^{9}	0.806 MeV ⎬	Gamma radiation
10^{-8}	1 Å		10^{8}	80.6 keV ⎭	
10^{-7}	10 Å		10^{7}	8.06 keV	Hard X rays
10^{-6}	100 Å		10^{6}	0.806 keV	Soft X rays
10^{-5}	1000 Å		10^{5}	⎫	XUV[a]
10^{-4}	10000 Å, 1 μ		10^{4}	⎬ Optical region	(See Table IIIa)
10^{-3}	10 μ		10^{3}		⎫ Infrared (IR)
10^{-2}	100 μ		10^{2}		⎬
10^{-1}	1 mm	300000 MHz	10	⎫	Millimeter
1	1 cm	30000 MHz	1	⎬	microwaves
10	10 cm	3000 MHz	10^{-1}	Radio astronomy	Centimeter microwaves
10^{2}	1 m	300 MHz	10^{-2}	region	Decimeter microwaves
10^{3}	10 m	30 MHz	10^{-3}	⎭	Meter microwaves
10^{4}	100 m	3 MHz	10^{-4}		Short waves
10^{5}	1 km	300 kHz	10^{-5}		Medium waves
10^{6}	10 km	30 kHz	10^{-6}		Long waves
10^{7}	100 km	3 kHz	10^{-7}		⎫ Very long waves or very low
10^{8}	1000 km	300 Hz	10^{-8}		⎬ frequencies (VLF)

[a] The spectral region between 1000 and 3000 Å is often called the XUV.

TABLE IIIa

λ (cm)	λ (customary units)	$s = 1/\lambda$ (cm^{-1})		Designation
1×10^{-5}	1000 Å; 0.1 μ	10^{5}	⎫	⎫ X, ultraviolet
2×10^{-5}	2000 Å; 0.2 μ	0.5×10^{5}	⎬ Ultra	⎬
3×10^{-5}	3000 Å; 0.3 μ	0.33×10^{5}	violet	
4×10^{-5}	4000 Å; 0.4 μ	0.25×10^{5}	⎭	Near ultraviolet
				Violet
5×10^{-5}	5000 Å; 0.5 μ	0.20×10^{5}		Blue
6×10^{-5}	6000 Å; 0.6 μ	0.167×10^{5}	Visible	Green
7×10^{-5}	7000 Å; 0.7 μ	0.143×10^{5}		Yellow
				Red
8×10^{-5}	8000 Å; 0.8 μ	0.125×10^{5}	⎫ Infra	Near infrared
9×10^{-5}	9000 Å; 0.9 μ	0.111×10^{5}	⎬ red	
10^{-4}	10000 Å; 1 μ	$0.10 \times 10^{5} = 10^{4}$	⎭	

The customary units in each wavelength region are printed in italics. The general designations of the wavelength regions are rather vague, and depend upon the method of observation. They have only a token significance.

We recall the following expressions relating the various units of energy:

$1 \text{ eV} = 1.601\,84 \times 10^{-12} \text{ erg};$

the energy of 1 eV corresponds to:

the wavelength $\lambda_0 = 12\,396.3 \times 10^{-8}$ cm,

the wavenumber $s_0 = 8067.1 \text{ cm}^{-1}$,

the frequency $\nu_0 = 2.41838 \times 10^{14} \, s^{-1}$,

the temperature $T = 11\,605.9$ K (using the formula $h\nu/kT = 1$),

or $T = 5040.4$ K (using the formula $h\nu/kT \ln 10 = 1$);

1 cm^{-1} corresponds to $1.985\,70 \times 10^{-16}$ erg.

2. The Opacity of the Atmosphere

Astronomical sources emit photons of all wavelengths; the distribution of these photons as a function of their energy constitutes the 'spectrum' of the source. The spectrum goes from gamma rays to radio waves and to 'VLF' (see Table III). It is more or less rich in radiation of various wavelengths, depending upon the nature of the physical emission processes. It is often represented by the functions F_ν or F_λ, which represent the flux emitted by the radiating surface – of the star – per second, per square centimeter, per frequency unit (or wavelength unit).

The terrestrial atmosphere reacts differently to radiation of different wavelengths; it also has its 'spectrum' – an absorption spectrum. We have seen the composition of the atmosphere: each of its constituents – molecular or atomic nitrogen; molecular, diatomic, or triatomic oxygen; carbon dioxide (CO_2), water (H_2O), etc. – has an absorption spectrum, and is not transparent at all wavelengths.

Figure 4 shows the reduced absorption spectrum of each of the principal gaseous constituents of the atmosphere – atomic nitrogen (N), atomic oxygen (O), molecular nitrogen (N_2), molecular oxygen (O_2), ozone (O_3), water vapor (H_2O), and carbon dioxide (CO_2): that is, the absorption coefficient k_{ν_i} per centimeter of thickness of the gas in question, which is assumed to be pure, at $T = 0°C$, and subjected to a pressure of one atmosphere (or 'exponential opacity per atmo-centimeter'). This reference volume corresponds to $N_0 = 2.688 \times 10^{19}$ molecules, a value easily derived from Avogadro's number.

We know the variation of N with altitude, for each atomic or molecular species (Figure 3). The 'optical thickness $\tau(h)$' of the atmosphere above the layer of height h is thus

$$\tau(h) = \int_h^\infty \sum_i k_{\nu_i} \frac{N_i(h)}{N_0} \, dh, \tag{1}$$

where the summation pertains to the various atomic and molecular species contained in the atmosphere. This formula defines the height h corresponding to given values of the optical thickness τ. This height obviously depends on the wavelength. The incident flux is reduced by 50% for $e^{-\tau} = \frac{1}{2}$, or:

$$\tau = \ln 2 = 0.693;$$

the corresponding height, $h_{1/2}$, is often considered characteristic of the altitude above which a vehicle must ascend, in order to observe radiation of the wavelength in question.

Figure 5 shows the function h for $\tau = 10^{-4}, 10^{-3}, 10^{-2}, 10^{-1}, 0.693, 1$, and 10. It is cut off at a wavelength of about 10 cm, because the different problems presented by the electrons and ions of the ionosphere interfere at long wavelengths, and are superimposed upon the true absorption (see Chapter IV, Section 3).

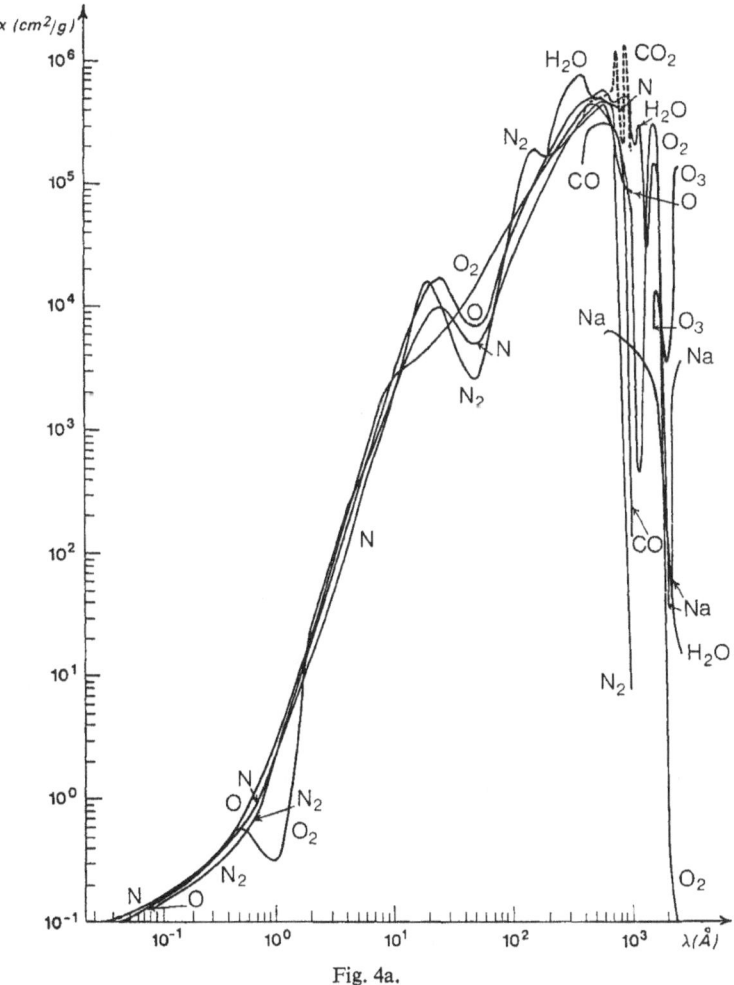

Fig. 4a.

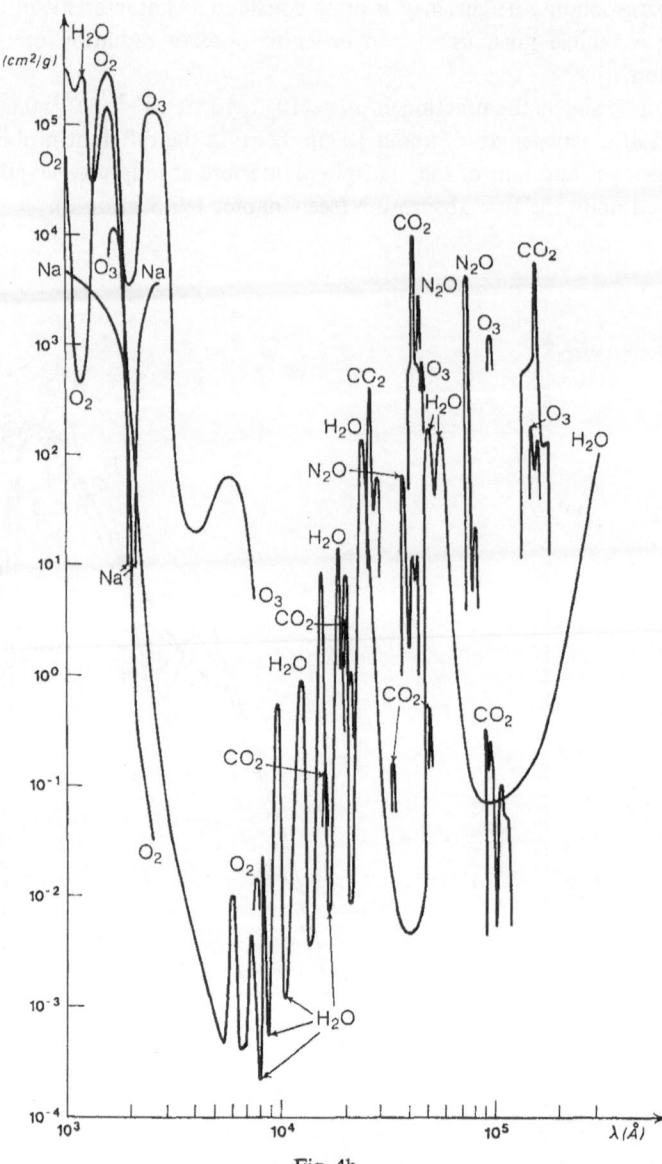

Fig. 4b.

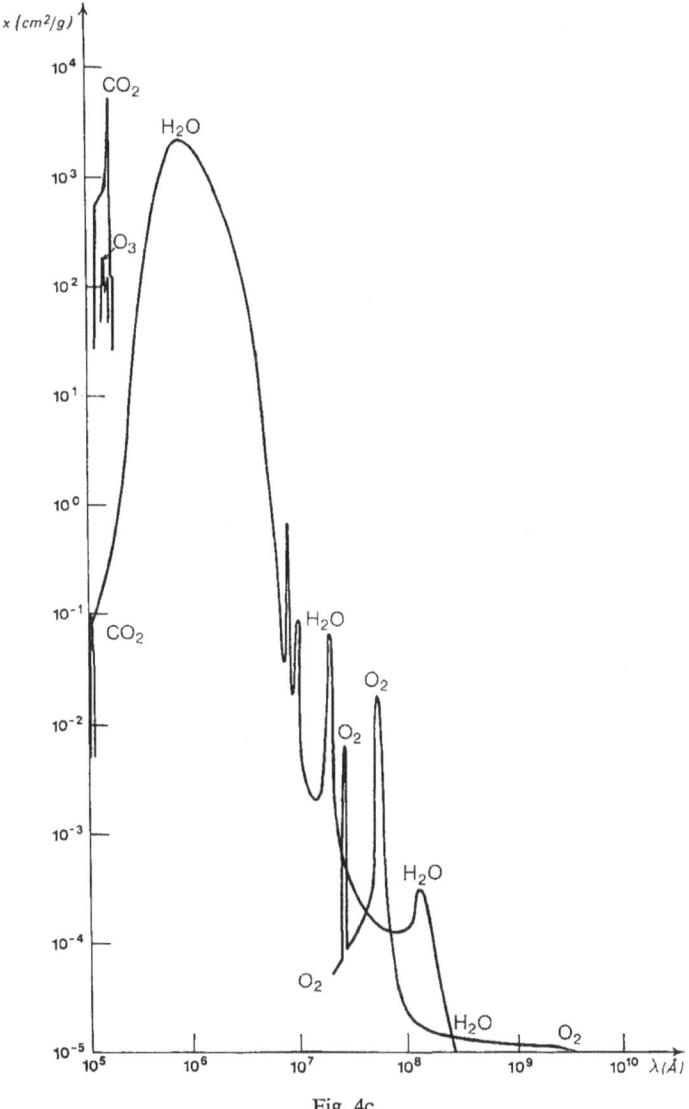

Fig. 4c.

Fig. 4. Absorption as a function of wavelength, for the various constituents of the atmosphere: (a) UV region; (b) visible and infrared; (c) radio region. – The scale of the *abscissa* is different in the three regions, but exactly comparable; *ordinate:* the 'reduced' absorption coefficient is expressed per cm of the gas, which is assumed to be pure, at $T = 0°C$, and under one atmosphere of pressure (the units of this quantity are cm²/g). – *Note*. Certain absorbents, especially ozone, are very sensitive to temperature and pressure, and consequently the 'reduced' coefficients used in this figure are not strictly correct; but the order of magnitude is not affected.

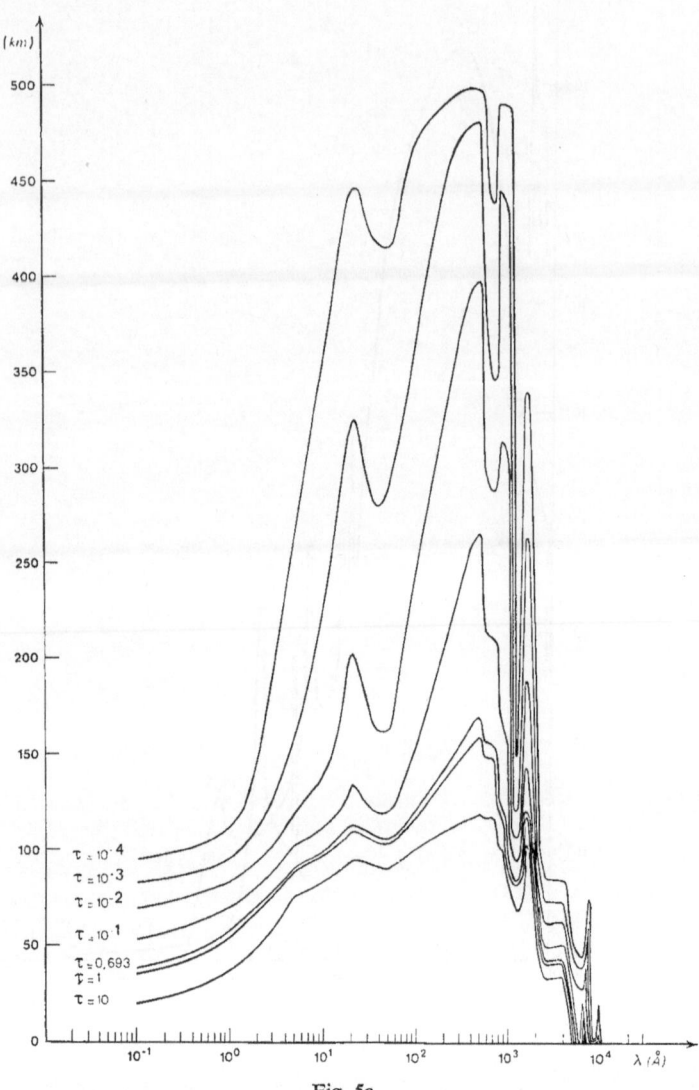

Fig. 5a.

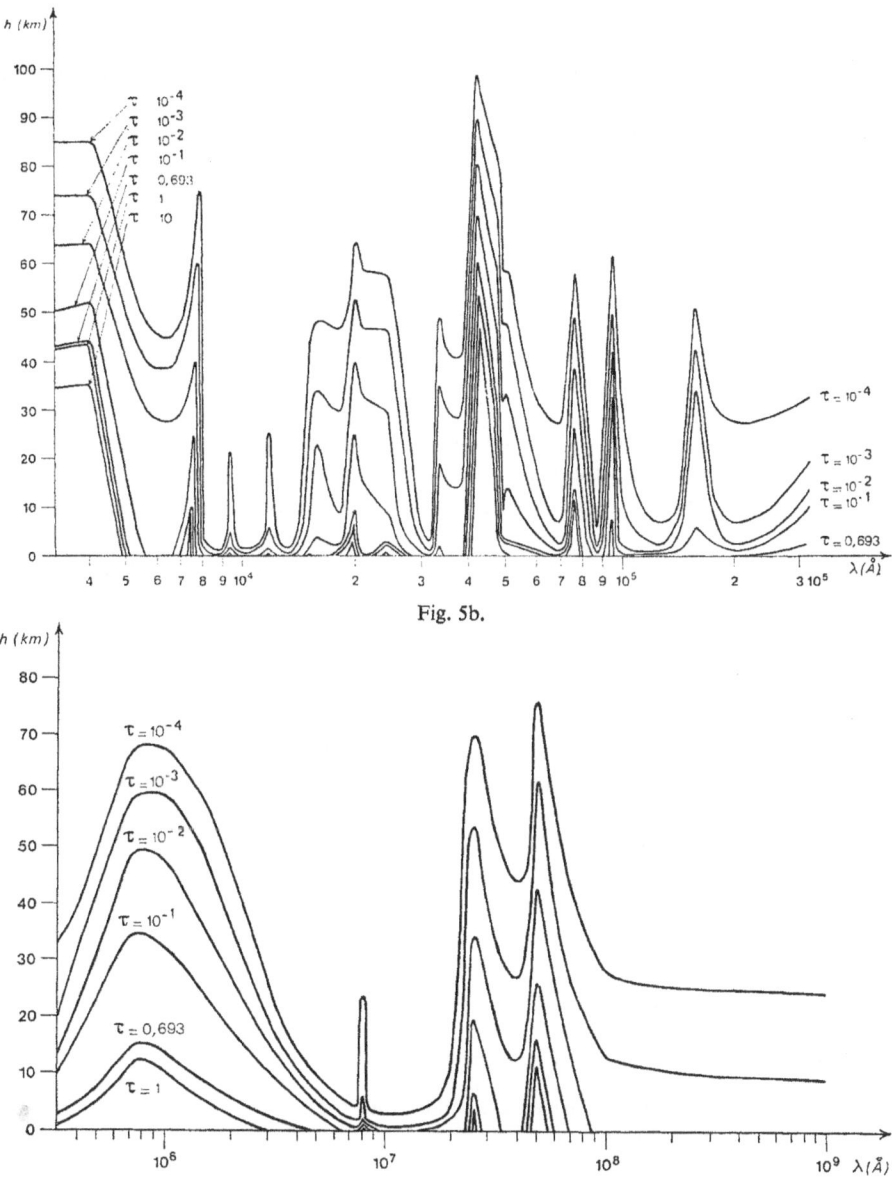

Fig. 5b.

Fig. 5c.

Fig. 5. Curves $h(\lambda)$ of equal optical depth τ, for different values of τ: (a) UV region; (b) visible and infrared; (c) radio region. – The scales (*abscissa and ordinate*) are different in the three regions; the abscissa scales are exactly comparable. The curve $\tau = 0.693$ (which corresponds to a reduction of the incident intensity from 1 to 0.5) more or less defines the altitude that must be attained for convenient observation of extra-terrestrial objects. – *Note*. Certain absorbents, especially ozone, are very sensitive to temperature and pressure, and consequently the 'reduced' coefficients used in this figure are not strictly correct; but the order of magnitude is not affected.

3. Atmospheric Absorption and the 'Windows'

We have seen in the preceding sections that the wall of the atmosphere is not equally opaque to all types of radiation. At certain wavelengths, its opacity is less – much less, even – than unity. Those regions of the spectrum to which the Earth's atmosphere is transparent ($\tau \ll 1$) are called 'windows'. The visible-near-infrared region, quite apparent in Figure 5, and the radio region (from a millimeter to a meter) are obvious examples. In the infrared, the water-vapor bands also have windows between them. These windows can be used from the ground by astronomers in traditional observatories, who look at the universe... through the window.

Note that the windows do not begin or end abruptly. On either side of a window, the opacity of the atmosphere changes slowly from a very small value to a larger value, corresponding to an optical thickness much greater than unity. Some windows are fully transparent, while others (shall we call them 'Venetian blinds'?) are insufficient for observations from the ground, but do make it possible to observe from mountains or in very dry regions, where water vapor does not affect the measurements.

Since the resources of terrestrial astronomy are still important, it will be useful to know these windows and blinds precisely, in order to explore them systematically. Table IV gives an approximate list of the most important spectral windows.

TABLE IV

The principal windows of the terrestrial atmosphere
at ground level

0.32 to 0.90 μ: near ultraviolet, visible, near infrared
(φ band of water)
0.95 to 1.1 μ: infrared
(Φ band of water)
1.15 to 1.3 μ: infrared
(ψ band of water)
1.5 to 1.75 μ: infrared
(Ω band of water)
2.0 to 2.4 μ: infrared
(χ band of water, carbon dioxide)
3.4 to 4.2 μ: infrared
(γ band: carbon dioxide CO_2, nitrous oxide N_2O)
4.6 to 4.8 μ: infrared
(water H_2O)
8.0 to 13 μ: infrared
(carbon dioxide CO_2)
16 to 18 μ: infrared
(water H_2O, oxygen O_2)
3 mm to about 10 m: radio astronomy
(ionosphere)

The principal absorbents which intervene between the
windows are indicated in parentheses.

We still have to know how to use the windows and blinds, and in particular how to correct the observations for the absorption, limited but not negligible, which is experienced by the radiation passing through them.

The rather vague concept of 'air mass' is often used to represent the atmospheric absorption at wavelengths for which $\tau < 1$. Using Equation (1), we can write:

$$\tau(h) = \bar{k} \int_h^\infty \rho(h)\,dh = \bar{k}\mathcal{M}. \tag{2}$$

In this expression, $\bar{k}$ represents the mean absorption coefficient of the air in the atmosphere; this approximation is known as 'Beer's law'. The air mass $\mathcal{M}$ can be evaluated without reference to the opacity, knowing only the variation $\rho(h)$ of the density of air with altitude. However, curvature must be taken into account in the calculation. If ζ is the angle between the line of sight and the vertical, we have

$$\mathcal{M} = \sec\zeta \int_0^\infty \frac{\rho(h)\,dh}{\sqrt{1 + (2h/R)\sec\zeta}} \sim \mathcal{M}_0 \sec\zeta, \tag{3}$$

a formula whose proof is purely geometric; $\mathcal{M}_0$ represents the air mass at the zenith of the place of observation, and R is the radius of the Earth.

There is also a classical procedure for reducing one's observations to outside the atmosphere, which consists of constructing a so-called 'Bouguer line' (Figure 6).

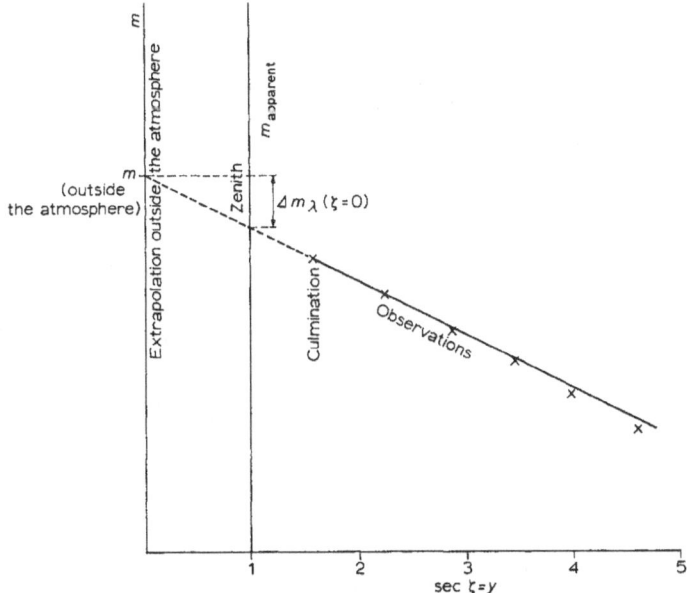

Fig. 6.   Construction of a Bouguer line. – *Abscissa:* $y = \sec\zeta$ specifies the direction of the observations, ζ being the angle between the line of sight and the vertical; *ordinate:* the magnitude of the object under investigation.

In a direction making an angle ζ with the vertical, the brightness of a source is reduced to:

$$E_{\mathrm{obs}}(\zeta) = E_* \, e^{-\tau(\zeta)}. \tag{4}$$

The quantity $E_{\mathrm{obs}}(\zeta)$ can be plotted as a function of ζ, which varies from $0°$ to $90°$; it is practical to take as the independent variable

$$y = 1/\cos \zeta = \sec \zeta.$$

Let us suppose the absorption is such that

$$\tau(\zeta) = \tau(\zeta_0) \sec \zeta$$

(this would be the case if the atmospheric layers were plane-parallel). Then we have

$$E_{\mathrm{obs}}(\zeta) = E_* e^{-\tau(\zeta_0) \sec \zeta} = E_* e^{-y\tau(\zeta_0)}, \tag{5}$$

or, in logarithmic form,

$$\ln E_{\mathrm{obs}}(\zeta) = \ln E_* - \tau(\zeta_0) \, y. \tag{6}$$

The 'magnitude' of a star is proportional to the logarithm of its brightness:

$$m_* = 2.5 \log_{10} E_* + \mathrm{const}. \tag{7}$$

Thus we have

$$m_{\mathrm{obs}} = m_* - y\tau(\zeta_0) \, (2.5/\ln 10), \tag{8}$$

or,

$$m_{\mathrm{obs}} = m_* - 1.09 \, y\tau(\zeta_0). \tag{9}$$

The variable $y = \sec \zeta$ varies from 1 (for $\zeta = 0$) to infinity (for $\zeta = 90°$). In this range ($y \geqslant 1$), we can plot the observed quantity m_{obs} as a function of y (Figure 6). In practice, the measurements may be greatly perturbed at the horizon; moreover, an astronomical object can culminate at large zenith distances, thus limiting the known range of the function m_{obs}. The theoretical line we would have been able to construct if Equation (6) were rigorously correct is often only moderately straight! In principal, however, extrapolating it to $y = 0$ should give us the value *outside the atmosphere* of the brightness E^* of the source, or its magnitude m^* (Figure 6).

This method has the great advantage that it avoids the use of tables for calculating the air mass, which is by nature an average value and does not take into account the conditions pertaining to a particular series of observations, made on a particular day and in a particular place.

But the use of this method is not always accurate, nor convenient. In the first place, the extrapolation of incomplete data is always a hazardous operation. And then, the method assumes that we can follow an astronomical object in its path across the sky – but along that path, the transparency of the sky is far from constant, either from place to place or as a function of time!

A difficulty of a different sort is the following: it is obvious that at the edge of a 'window', τ varies rather rapidly with wavelength. Consequently, an error in the wavelength will produce an error in the construction of the Bouguer line. Moreover, the geometric irregularities of the absorbents (water vapor, for example) add to the uncertainty. Herein lies, no doubt, the cause of the widely differing values obtained by different authors for the infrared flux from the center of the solar disk.

CHAPTER III

ATMOSPHERIC DIFFUSION:
EXTINCTION AND THE BLUE OF THE SKY

We have seen that the atoms and molecules of air *absorb* incident radiation at certain wavelengths. The energy absorbed is transformed into heat (and radiated by the atmosphere at other wavelengths, in the infrared). But air can also *diffuse* light, without absorbing it, by redistributing the incident energy in all directions. This is 'Rayleigh scattering' by air molecules. It has a double effect which has been known, at least qualitatively, since the time of Leonardo da Vinci – an effect that manifests itself in all the spectral windows, but particularly in the visible region.

First, it eliminates an important fraction of the direct radiation from astronomical sources.

Then, it scatters in all directions the radiation it has removed.

Let us examine these two effects in turn.

1. Atmospheric Absorption in the Visible 'Window'

Rayleigh scattering removes an important fraction of the solar or stellar radiation, a fraction that increases as the wavelength becomes shorter. Figure 7 gives this extinction as a function of wavelength, at ground level and at the zenith. Diffusion alone is considered, for the absorption is smaller. We see that air alone can dim the light of the stars through Rayleigh scattering. Clearly, if dust particles or droplets (clouds, soot, etc.) accumulate above the astronomer, the diffusion will be even greater. The solid curve in Figure 7 takes into account only the best possible conditions at ground level.

Let Δm_λ be the gain in magnitude, at a given wavelength, that would be obtained by eliminating the atmosphere above an observer at altitude h_0. We have

$$\Delta m_\lambda = 2.5 \log_{10}(E/E_0) = 1.09 \, \tau_{v_0} \sim \tau_{v_0}, \tag{10}$$

where τ_{v_0} is the optical thickness of the atmosphere above altitude h_0, and $1/\log_{10} e = 2.302$ has been set ≈ 2.5.

Although the nature of extinction by absorption is different from that produced by Rayleigh scattering, these two effects add arithmetically in the calculation of air masses and opacities. We shall consider, in practice, only the sum of the two – thus the method of the Bouguer line measures the combined effect of absorption and diffusion. Figure 7 shows the relative importance of these different physical sources of opacity, in the visible window.

The effect of Rayleigh scattering is dominant. It is clear that the magnitude loss

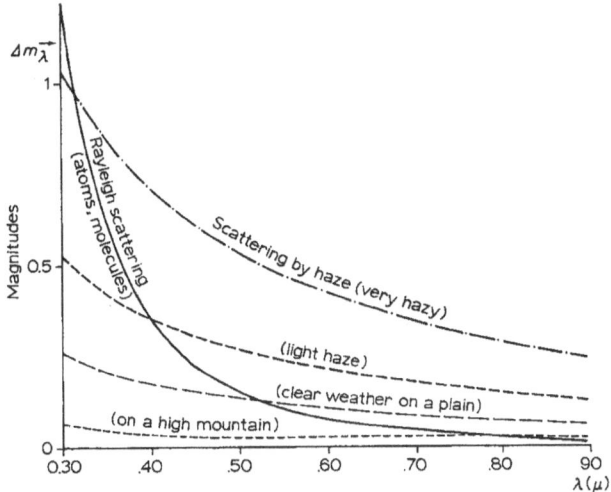

Fig. 7. Attenuation of the radiation of celestial bodies by atmospheric diffusion. – *Solid line:* effect of Rayleigh scattering by air molecules on the brightness of a source at the zenith. The *dashed* lines assume that the scattering effect is proportional to $\beta\lambda^{-\alpha}$, where $\alpha = 1.3$ and β takes on four different values corresponding to haze of different thicknesses.

Δm_λ depends strongly on the wavelength, as does the optical depth of the atmosphere. Since the Rayleigh law varies as λ^{-4}, we can deduce that it would be useful to work from a high mountain in order to observe the ultraviolet portion of the spectrum. This is one of the reasons for the creation of the high-altitude station at Jungfraujoch.

This phenomenon, somewhat less pronounced, also exists if there is haze. Figure 7 describes dimming by haze in a few typical cases.

These few arguments clearly show how useful it would be, even in the spectral windows, to by-pass the atmosphere through space research. Such experiments would avoid the complications and uncertainties of the different corrections which arise in the comparison of observations – a necessary step in the widely-used methods of differential spectrophotometry (or ordinary photometry).

But these considerations should not make us overlook the fact that, for differential measurements, it is not really the weakening of the radiation from a certain star that is troublesome – it is rather the fact that the contrast with the sky background often becomes too small. But the sky background is also produced by Rayleigh scattering. And it is with this problem that we shall now concern ourselves.

2. The Blue of the Sky

Coupled with extinction, Rayleigh scattering is an essential phenomenon. It super-imposes upon the direct radiation from a given astronomical source, a diffuse radiation derived from the light of all the other heavenly bodies – radiation that is intense by day and weak by night, hiding faint objects and imposing a limiting magnitude on observable objects. By selecting out the short-wavelength radiation, Rayleigh scattering is responsible for the blue color of the diffuse radiation – in other words, for the 'blue of the sky'.

Can we evaluate the limiting magnitude for observation, taking Rayleigh scattering into account?

3. Sky Brightness in the Light of the Sun or Full Moon

To simplify the argument, we shall consider first the case of a single illuminating body, the Sun or full Moon. The incident luminous intensity, I_0, arising from each point of the source, is attenuated by the atmosphere on account of Rayleigh scattering. The observed intensity is thus

$$I_0 e^{-\tau \sec \phi},$$

where τ is the optical depth corresponding to Rayleigh scattering (the other sources of opacity are assumed to be negligible in the wavelength region of classical astronomy, between 0.3 and 10 μ), and φ is the zenith distance of the object. The total diffuse intensity is therefore

$$I_0 (1 - e^{-\tau \sec \phi}).$$

The flux F_0 of diffuse energy is equal to the product of this quantity with the apparent surface area of the source – that is, with $\pi \omega^2/4$, where ω is the apparent diameter of the object. (For the Sun and Moon, $\pi \omega^2/4 \approx 6.8 \times 10^{-5}$ steradian.)

We know that the scattering is not isotropic, and that the diffuse radiative flux is not uniformly distributed over the sky. In order to evaluate the limiting magnitude correctly, we must therefore estimate the sky brightness in each direction. Its intensity I will obviously be proportional to the scattered flux: $F_0' = F_0 (1 - e^{-\tau \sec \varphi})$. At each point of the sky, the intensity depends on several parameters:

(a) the direction of the Sun, defined by its angular zenith distance θ_0;

(b) the direction of observation, defined by its angular zenith distance θ;

(c) the altitude h, which essentially determines the opacity parameter τ;

(d) the transparency of the air, which also governs τ;

(e) the wavelength, upon which the Rayleigh scattering coefficient depends, again influencing τ.

In general, we shall assume the best possible conditions – that is, a truly blue sky, unwhitened by haze. The only scattering agents are assumed to be air molecules.

Under these conditions, we can calculate the distribution of $I(\theta, \theta_0)$, expressed in terms of the incident flux F_0'. Such a calculation rests upon the theory of radiative transfer in the terrestrial atmosphere; it must take into account not only primary scattering, but also secondary scattering and anisotropy. Figure 8 gives an example of the results of such a calculation for a particular case, according to Chandrasekhar. It is out of the question for us to explain his method in any detail, for it involves very elaborate mathematics.

We see that I/F_0 (the ordinate) is on the order of 10^{-2} to 10^{-1}, except at the horizon; therefore I will be on the order of $10^{-6} I_0$. This result depends on the wavelength, through τ. In Figure 8, it was assumed that $\tau=0.2$; this roughly corresponds to a wavelength of 4500 Å, according to Figure 7 (since Δm and τ are of the same order of magnitude). If the wavelength is larger, τ obviously decreases; consequently, it is better to work whenever possible in the red or infrared, if we want to minimize scattering by atoms and molecules in the atmosphere.

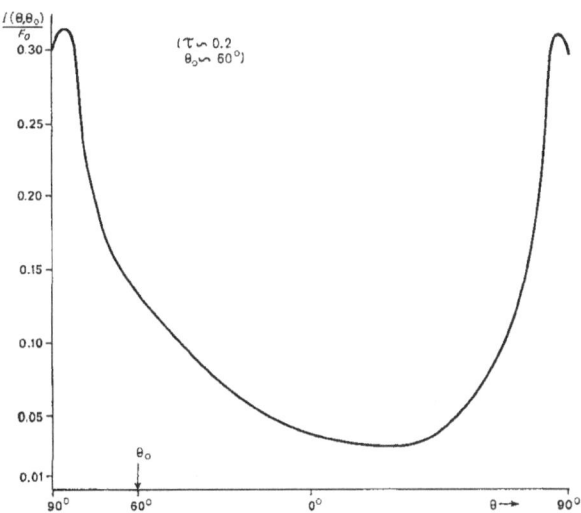

Fig. 8.   Distribution on the sky of diffused sunlight (after Chandrasekhar). – *Ordinate:* the intensity; *abscissa:* angle between the normal and the line of sight. This figure is an example of the solution of the equation of radiative transfer in a planetary atmosphere, whose optical thickness at the chosen wavelength, as well as the direction of the Sun, is a parameter of the calculation.

We can, then, evaluate the magnitude of the sky *per unit solid angle*; let m_{sky} be this magnitude. To make a rough calculation, let us assume that I is equal to $10^{-6} I_0$; we find then that for the Sun and Moon (which have the same apparent size), we have

$$m_{sky} = m_\odot + 4.5 \quad \text{(Sun at the zenith)}, \tag{11}$$

$$m_{sky} = m_\mathbb{C} + 4.5 \quad \text{(full Moon at the zenith)}. \tag{12}$$

To be completely rigorous, this expression should, of course, depend on θ, θ_0, and τ. Now, we know the magnitudes of the Sun and Moon. If we take the photovisual region of the spectrum for reference, we have

$$m_\odot = -26.9 \quad \text{and} \quad m_\mathbb{C} = -12.7. \tag{13}$$

In the photographic region, we have

$$m_\odot = -26.4 \quad \text{and} \quad m_\mathbb{C} = -12.2. \tag{14}$$

What is interesting to know is not, of course, the magnitude per unit solid angle, but the magnitude of a portion of the sky corresponding to the apparent surface area of a star. For a sufficiently large instrument, the vibration of the images fixes their apparent dimensions; then we can assume that the diameter of the image is $\alpha = 1''$ under average seeing conditions, and $\alpha = 0''.2$ under very good conditions.

The magnitude of an area of the sky equal to the area of a stellar image is, for these two cases:

	$\alpha = 0''.2$	$\alpha = 1''$
Sun,	$m_{sky} = + 7.9$	$+ 4.4$
Full Moon	$m_{sky} = + 22.1$	$+ 18.6$

But although these numbers are an indication of the limiting magnitude, they do *not* give its actual value; we shall see in the following sections why this is so.

4. Sky Brightness on a Moonless Night

On a moonless night, the sky is still illuminated – very weakly, but sufficiently to be a limiting factor on the performance of large instruments. Space research will free us from some of the sources of this extraneous light. It is therefore important to examine them in more detail.

The sources of extraneous radiation are as follows:

I. Extraterrestrial objects: a large number of *unresolved* stars, galaxies and nebulae certainly contribute to the sky background.

II. The light of these extra-terrestrial sources (resolved or otherwise) diffused by the atmosphere.

III. The light diffused by the interplanetary medium (especially by the outer regions of the Earth's atmosphere, the 'geomagnetic tail'; and by the zodiacal light, the dusty extension of the solar corona).

IV. The light emitted by the terrestrial atmosphere itself – aurorae, night-sky luminescence, and even thermal radiation at short wavelengths.

V. Finally, artificial light (from cities) diffused by the atmosphere.

Component V can obviously be avoided by a judicious choice of the observatory site; in the following discussion, we shall assume that this choice has, in fact, been so judicious as to eliminate it completely.

Components I and III are always present; space research will not be able to eliminate them. Component II adds nothing to Component I as far as the sky as a whole is concerned, but it smoothes out the fluctuations from one point on the sky to another.

Component I is estimated (from counts of astronomical objects) at about -2.4 magnitudes per unit solid angle, in the darkest regions of the sky. In an apparent surface area of diameter $\alpha = 1''$, its magnitude is therefore 20.4 (a rather optimistic estimate); if $\alpha = 0''.2$, the magnitude is 23.9. This value obviously varies from one point on the sky to another, as is indicated in more detail in Table VI.

Component IV is important, but less so than Component I. It is on the order of $m=23$ per square second, in the visible region. Its spectrum, instead of being continuous, is essentially a line and band spectrum; the continuous component is very poorly known. We shall assume that it can be satisfactorily reduced by means of well-chosen filters. But in any case, space research will do an even better job of eliminating it. Table V lists the wavelengths of the most important of these lines and bands, and their respective intensities.

Table VI summarizes the preceding discussion by providing the order of magnitude of the different components of the night-sky brightness, in the photographic and photovisual regions.

TABLE V

Element	Wavelength (Å)	Intensity (relative units)	Emission height (very poorly known) (km)
O I	5577	260	250
O I	6300.6364	190	400
Na I	5890.5896	150	200
N_2 (Vegard-Kaplan bands)	4000 to 5000	100	200
O_2 (Hertzberg bands)	3000 to 4000	100	70 to 100
OH (rotation-vibration bands)	0.6 to 1 μ 1 to 4.4 μ	10^{5} [a] 10^{6} [a]	} 70 to 100

[a] The emission of these last two groups of bands between 0.8 and 2.2 μ is greater than the stellar and interplanetary background.

TABLE VI

Source	Photographic region		Photovisual region	
	(1)	(2)	(1)	(2)
(a) Faint stars ($m > 6$):				
galactic pole	16	-1.83	30	-2.49
average area of the sky	48	-2.99	95	-3.74
galactic equator	140	-3.17	320	-5.07
(b) Additional galactic radiation, and light from bright stars, diffused by the atmosphere	20	-2.04	40	-2.79
(c) Zodiacal light (far from the plane of the ecliptic).	20	-2.04	30	-2.49
(d) Noctiluminescence (at the zenith):				
atomic lines	0		50	-3.04
resolved molecular bands	30	-2.49	60	-3.24
continuum and unresolved bands	70	-3.39	150	-4.24
Total (average area of the sky, at the zenith)	200	-4.54	400	-5.29
Total (outside the atmosphere)	100	-3.79	240	-4.74
Total (far from the solar system)	80	-3.54	210	-4.59

Column (1) gives the equivalent of the sky background in number of tenth-magnitude solar-type stars per square degree. Column (2) gives the magnitude of the sky per steradian. The sources that could be eliminated by space research are printed in *italics*. But as far as the zodiacal light is concerned, this achievement is clearly a long way off!

5. The Limiting Magnitude for Detection of a Star

(under normal observing conditions)

In the circumstances we have just described, how can we define the limiting magnitude for detection of a star or, in general, of an extended object of apparent diameter β which is assumed to be uniformly illuminated?

The problem is far from simple. It obviously depends on the detector. This detector may be insufficiently sensitive to the light of the star; it may, on the other hand, be so sensitive that it is saturated by background light and cannot detect any additional source. The result also depends on the *telescope* – this is a well-known experimental fact.

The difficulty arises in that the beam of light (or, often, the ray of light – for we are interested in looking for faint sources and in minimizing the extraneous light, making it no more troublesome than necessary) is not constant, or even continuous. If it were, any signal would be detectable (between the threshold and the saturation point of the receiver), however great the background noise. The only limit would be imposed by the sensitivity of the receiver. In reality, what is important for any receiving apparatus (the eye, the photographic plate, the photocell or image tube) – what is *measurable* in a given experiment – is not so much the intensity of the radiation or the number of photons received, but the number of photons actually counted – the number of *effective* photo-events; and this number varies with time. If n is the number of photons per second emitted by the light source and reaching the detection system; t the duration of the experiment; and q a number less than unity which measures the efficiency of the detector for the photons in question – that is, the fraction of photons capable of including effective photo-events; then the number of photo-events corresponding to one measurement is

$$v = nqt. \tag{15}$$

This number fluctuates from one experiment to the next: the arrival and selection of photons obey the 'laws of chance'. It can be shown that if the experiment is repeated a large number of times, the successive measurements of v will have a certain dispersion, and the 'standard' dispersion will be (Figure 9):

$$\Delta v \sim v^{1/2}. \tag{16}$$

In other words, repeating a measurement that corresponds to four photo-events will give a result that can be written: 4 ± 2. For 100 photo-events, this becomes 100 ± 10; and for 10000 photo-events, 10000 ± 100. The relative error decreases as the number of photo-events in the experiment increases.

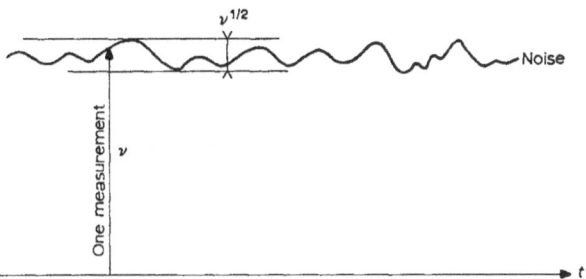

Fig. 9.   Fluctuations in a photon-counting procedure.

The background radiation that falls upon a detector from the sky obeys this law. A star can be detected only if the number of photo-events corresponding to its radiation in a given experiment exceeds – not the number v of extraneous photo-events – but a number of such events of order of magnitude $\sqrt{v}$ over the same area of the sky.

The threshold of detectability is therefore defined by

$$Nqt \geqslant k(\alpha^2\, nqt)^{1/2}, \tag{17}$$

where k is a number of order unity, α is the apparent diameter of the stellar image, and n is the number of background photons per unit solid angle. In this calculation we have approximated $\pi/4$ by 1.

If we use a telescope of diameter D, since N and n correspond to numbers of photons per unit area of a receiver placed at the focus of the instrument, we can write:

$$D^2\, Nqt \geqslant k(\alpha^2\, D^2\, nqt)^{1/2}. \tag{18}$$

Since there may be background radiation from the instrument as well as from the sky, we must write

$$D^2\, Nqt \geqslant k\alpha(D^2\, nqt + \mathcal{N}t)^{1/2} = k\alpha D(nqt)^{1/2}(1 + \delta)^{1/2}, \tag{19}$$

where the coefficient δ represents the extraneous contribution of the instrument.

For a star at the threshold of detectability, the ratio of the stellar intensity to the background noise will be:

$$y_0 = \frac{D^2 Nqt}{\alpha^2(D^2 nqt + \mathcal{N}t)} = \frac{k}{\alpha(D^2 nqt + \mathcal{N}t)^{1/2}} = \frac{k}{D(nqt)^{1/2}(1 + \delta)^{1/2}}. \tag{20}$$

If m is the magnitude of the sky per steradian (unit solid angle), and m_0 the magnitude of a star at the threshold of detectability, we can easily calculate m_0. For the magnitude of the sky, referred to the apparent surface area of the star, is

$$m_\alpha = m - 2.5 \log(\alpha^2) = m - 5 \log\alpha, \tag{21}$$

where α is, of course, expressed in radians.

Thus we have

$$m_0 - m_\alpha = -2.5 \log(N/n\alpha^2) = -2.5 \log(y_0/(1+\delta)), \tag{22}$$

or,

$$m_0 = m - 2.5 \log\alpha - 2.5 \log k + 1.25 \log t + 3.75 \log(1+\delta)$$
$$+ 1.25 \log q + 1.25 \log n + 2.5 \log D. \tag{23}$$

We can rewrite this expression, making use of the fact that m is proportional to $2.5 \log n$:

$$m = -2.5 \log n + 2K. \tag{24}$$

The constant K can be calculated in such a way as to link this formula with the usual magnitude systems; but we shall not derive its value in the following sections.

Thus we have:

$m_0 = 0.5\, m\ -2.5 \log\alpha$ $\begin{cases} \text{determined by the dimensions of the instrument, if it is small;} \\ \text{determined by the seeing, for large instruments; determined} \\ \text{by the dimensions of the source, if it is extended;} \end{cases}$

$\begin{aligned} &+ 1.25 \log t \\ &+ 1.25 \log q \end{aligned} \Big\}$ conditions of the experiment;

integration time, efficiency

$+ 2.5 \log D$ instrument chosen

$+ 3.75 \log(1+\delta)$ background noise within the instrument

$+ 2.5 \log k$ safety factor

$+ \text{constant } K$ $\begin{cases} = 7.65 \text{ (pg system)} \\ = 7.22 \text{ (pv system)} \end{cases}$ (25)

This expression reveals a certain number of remarkable facts.

First of all, we note that m_0 depends upon $0.5\, m$, and not upon m. This is essentially due to the introduction of Equation (24), which is applicable only if we are far enough from both the saturation point and the threshold of the instrument.

Next we see that m_0 depends upon D through the term $2.5 \log D$, and not $5 \log D$. This relation has been verified experimentally to a high degree of accuracy.

Finally, we see in the term $2.5 \log\alpha$ the importance of the image quality for the limiting magnitude. For a small instrument, α is on the order of $1.22\, \lambda(f/D)$ for visual observations, and on the order of the size of the photographic grains – if this is greater than $1.22\lambda(f/D)$ – for photographic observations. For a larger instrument, α is fixed by the atmospheric turbulence alone ($\alpha = t$; see below, Chapter IV, Section 4). In the case of nebulae, α obviously corresponds to their apparent dimensions.

Equation (25) is completely valid in practice, if observations are made with the eye, photoelectric photometry, image tubes below their saturation point, or normally exposed photographs.

Take, for example, the eye. 'Reasonable' estimates of the various quantities are:

$$m = -5.3 \quad \text{(see Table VI)}; \tag{26}$$

$q \sim 0.02$; $t \sim 0.02$ sec; $\alpha \sim 6 \times 10^{-4}$ radian (2'); $D \sim 0.4$ cm; $k \sim 5$; $\delta \sim 1$. These values lead to a threshold for visual observation of

$$m_{vis} = 6.02 \text{ magnitudes}. \tag{27}$$

We know that visual observation reaches only the sixth magnitude – this is the very basis for the definition of the magnitude scale! The agreement is therefore excellent. But let us be frank – the author has slightly manipulated the 'reasonable' estimates above! The product qt could be two or three times greater, α smaller, D a little larger; the estimation of k is entirely arbitrary. The thing to remember about this calculation is that it gives the right order of magnitude; but it should be used only for deductions of a differential nature.

6. Detection near the Saturation Point of the Receiver

What happens in the limiting case, when one tries to 'bring out' faint objects by over-exposing? We know from experience that such conditions are not amenable to exact calculation. But what are the limiting magnitudes that can be attained? In this case, the number of photo-events of noise no longer depends on t or q or D – it is determined only by the behavior of the receiver at saturation. We have

$$v = \alpha^2 S \quad \text{(where S is a constant characteristic of the receiver)}, \tag{28}$$

and consequently:

$$m_0 = m - 2.5 \log \alpha - 2.5 \log k + \text{constant}. \tag{29}$$

Thus m_0 is proportional to m (and not to $0.5\,m$, as in the unsaturated case). But the effect of atmospheric turbulence remains the same.

There has naturally been a great deal of discussion concerning the importance of the instrument and, especially, the properties of detectors. We shall base our argument upon an empirical value, which is in agreement with the above theory; and we shall try to see to what extent space research facilitates the study of very faint objects, by increasing the limiting magnitude.

To begin with, the limiting magnitude under extreme conditions (saturation) of photographic observations, for the largest existing instrument on Earth (Mt. Palomar), is $m_0 \sim 23.5$ (an empirical determination). Table VI shows that by space research we can gain at most one magnitude in the magnitude of the sky (whatever the area over which this magnitude is estimated), and therefore in m_0.

On the other hand, the elimination of seeing gives us, instead of $\alpha \sim t$ (for example), $\alpha \sim 1.22\lambda/D$. If D is large enough (and this means that the focal length must also be fairly large, since there are obvious limitations on the f-ratio), α can be very small. Thus if $D \sim 10^2$ cm, α is on the order of $0''.1$; if $D \sim 10^3$ cm, then α is on the order of $0''.01$. In this fashion one gains two to five magnitudes in comparison with normal conditions in the terrestrial atmosphere. We shall see how space research may one day be used in order to take advantage of this very important gain.

7. The Study of Moderately Bright Objects

It is not only faint objects whose investigation can be made easier. We have seen that around the Sun, the sky is rather bright; around the full Moon also. We shall return later to the type of program which requires the elimination of this background light. But in this case there is no need to approach the saturation point; and Equation (25) will give us the best information concerning the gains to be anticipated.

Thus by going outside the terrestrial atmosphere we can gain about 18 magnitudes (using the data in Section 3) with respect to the sunlit sky, and four magnitudes for a sky with full Moon. To this improvement, we can add the few magnitudes to be gained by eliminating turbulence and by using large instruments; we shall see in Part II, Chapter II what programs are best adapted to these possible gains.

In the same spirit, we would do well to recall (with an eye to observations of the solar corona) that the intensity of the sunlit sky was taken in Section 3 to be $10^{-6} I_0$. The limiting magnitude (per steradian) near the Sun, for the study of the corona, depends upon the observing conditions. In practice, with existing instruments, coronal intensities equal to $10^{-7} I_0$ have been detected without much difficulty. Gaining 18 magnitudes would then be equivalent to reaching intensities of $10^{-14} I_0$, 10^8 times fainter than the sky as observed from the ground.

Note that the advantage of space research is more apparent for daylight observations than for nocturnal observations. It is possible, then, that an important advantage could be reaped even without attaining large altitudes. The calculation could be made as follows:

At an altitude of observation h, the light absorbed – and therefore re-diffused – is $\tau/\tau_0 = e^{-h/H}$ times that absorbed at ground level, neglecting secondary scattering by the atmosphere beneath the observer and assuming $\tau < \tau_0 \ll 1$ (which is correct for $\lambda > 4000$ Å).

In any given direction, the gain in 'blackness' of the sky, in magnitudes, is

$$\Delta m \simeq 2.5 \log_{10}(e^{-h/H}) \sim 1.08\,(h/H). \tag{30}$$

In this calculation, we assume Rayleigh scattering by a purely molecular and atomic atmosphere (the absorption is proportional *only* to the absorbing mass), and an exponential density distribution (see above).

We see, then, that if $H \sim 8$ km we gain about one magnitude on a very high mountain, and three magnitudes in a balloon. This would enable us to attain coronal intensities on the order of $10^{-8} I_0$, perhaps even $10^{-9} I_0$.

According to this calculation, the gain does not depend on the wavelength. This would not be true if we took haze into account, for its altitude distribution is different from that of the atoms and molecules responsible for Rayleigh scattering. Naturally, τ becomes negligible at satellite altitudes, and the preceding calculation is no longer valid. Since conditions for ground-based observations are worse at short wavelengths, it is at these wavelengths that the advantage of satellites will be most obvious.

CHAPTER IV

THE INHOMOGENEITY OF THE EARTH'S ATMOSPHERE: CLOUDS AND REFRACTION

1. Meteorological Conditions

The reader will excuse me if I remind him that observatories on the ground (and even on high mountains) are greatly inconvenienced by meteorological conditions.

First of all, it is clear that clouds completely prevent observation. Having made this categorical statement, we must add one qualification: radio astronomers do, in fact, observe through the clouds, and radio astronomy is an important extension of astronomy. But it cannot do everything, and to be convinced of this it suffices to behold the ardor with which observatory builders seek out sites devoid of cloud covering (insofar as possible). They go to Arizona, Anacapri, central Turkey, Baja California...

But there are perhaps worse things than not being able to observe at all. To be sure, when the sky is overcast night after night and the astronomer, on an expedition to Pic-du-Midi, has to fall back on television or bridge, his morale quickly sinks very low. But what about the man who has been working all night at the telescope and who, exhausted, looks over his readings by the wan morning light (an astronomer's morning light is always wan), only to find that they are useless because of the continual, unsuspected passage of cirrus?

For we must recall that clouds do not always form an infinitely opaque screen. There are, of course, very low strato-cumulus, which clearly prohibit all observations. But there are also light cirrus, at altitudes ranging from 5 to 13 km, which can be seen in broad daylight as whitish floss on a blue-tinged sky – but which remain transparent, and can be detected by night only by means of the spurious fluctuations they produce in the brightness of the stars. It is not unusual for passing cirrus to introduce fluctuations of several tenths of a magnitude. And we want measurements made to a hundredth of a magnitude! The spectroscopists generally do not care – they integrate over time, and expose their plates a little longer if their colleague the photometrist tells them that cirrus is present. But what is the photometrist himself to do? He is trying to find variations in the light of certain stars, explosions in the life of stellar embryos, who knows what? But all he can really detect is the insidious passage of cirrus. We say, then, that the sky is not 'photometric'. It would be a mistake to neglect this annoying aspect of ground-based observations – for since the cirrus clouds are at high altitudes, the sky will cease to be photometric simultaneously over an extended region. Although the micro-climate, as we shall see, can influence certain sites very favorably or very unfavorably by the degree of atmospheric turbulence, which varies rapidly from point to point, it is not so easy to choose a site whose photometric qualities are certain – even on a high mountain!

But space astronomy will clearly be free of such inconveniences.

Another important aspect of meteorological conditions, for the astronomer, is the wind pattern, especially when there are temperature inhomogeneities. But to explain this point, we shall first have to talk about atmospheric refraction.

2. Atmospheric Refraction

Whatever its wavelength, in the visible as well as in the radio region, the radiation coming from astronomical sources is refracted by the air. This refraction can be described as the superposition of a *mean* refraction, produced by the perfectly horizontal layers, and an *aleatory* refraction due to the motion of the air layers and thus to the irregular structure of the refracting medium. It is almost obvious that the mean refraction will be very important at the horizon and negligible at the zenith; the same cannot be said of the aleatory refraction which, although it is indeed more important at the horizon, can also be very important at the zenith. The first type of refraction affects the mean position of the images, the second their stability – and therefore their quality. Both effects are very troublesome in ordinary astronomy, and severely limit its potential.

The physical quantity of interest here is obviously the index of refraction of air, which is associated with its specific mass. This index (according to the measurements of Barrell, 1961) can be represented in the visible region by the formula

$$10^{+6}(n_0 - 1) = 2.727\,29 \times 10^2 + \frac{1.4823}{\lambda_0^2 \times 10^8} + \frac{2.041}{\lambda_0^4 \times 10^{16}} \times 10^{-2}, \qquad (31)$$

where λ_0 is expressed in centimeters (so that $10^4 \lambda_0$ is the wavelength in microns). This formula is, of course, valid only under certain well-defined conditions: a pressure $p = 76$ cm of mercury, a temperature of $15\,°C$, and zero water-vapor pressure ('dry' air).

The variation of n with temperature and pressure can, however, be calculated by means of the formula

$$(n - 1) = (n_0 - 1)\,\frac{p\,[1 + (1.049 - 0.0157\,t)\,10^{-6}\,p]}{720.883\,(1 + 0.003\,661\,t)}, \qquad (32)$$

where p is expressed in centimeters Hg and t in $°C$. If we assume that water vapor is present at a partial pressure p_{H_2O}, we have

$$(n - 1) = -(n_{dry} - 1)\,\frac{0.0624 - 0.000\,680/(\lambda_0^2 \times 10^8)}{1 + 0.003\,661\,t}\,p_{H_2O}, \qquad (33)$$

where λ_0 is expressed in centimeters (note the minus sign!).

Figure 10 shows the variation of $(n - 1)$ in the visible region and the near infrared, with $t = 0\,°C$, $p = 760$ mm Hg, and $p_{H_2O} = 4$ mm Hg. For other temperatures and pressures, one has only to multiply by the quantity

$$y = \frac{p}{760 + 2.9\,t},$$ (34)

which accounts rather well for the change in water-vapour pressure with temperature, and assumes simply that the air has a mean composition.

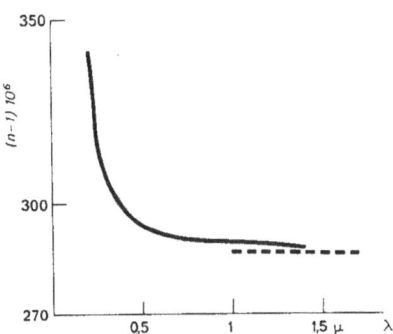

Fig. 10.   Refractive index of air.

In the radio region, the refractive index takes on values very close to those in the visible region. For wavelengths of less than 1 m, it is practically independent of wavelength and can be sufficiently well represented by the formula:

$$(n-1) \times 10^6 = 288\,\frac{p}{760(1 + 0.003\,661\,t)} + \frac{6.7\,p_{H_2O}}{(1 + 0.003\,661\,t)^2}.$$ (35)

For $p = 760$ mm Hg, $t = 0\,°C$, and $p_{H_2O} = 10$ mm Hg, for example, we have

$$(n-1) = 0.000\,355\,5,$$ (36)

where the term due to water vapor has become very important (and positive – whereas in the visible region it appeared with a minus sign). Figure 11 enables us to compare the optical refraction (at 5200 Å) with the radio refraction (decimeter and centimeter wavelengths).

Knowing the value of the index n and its variation with altitude h (through the functions $t(h)$ and $p(h)$), it is not difficult to calculate the refraction correction – that is, the deviation of light rays which have passed through the atmosphere. We can write

$$\zeta_\infty = \zeta_0 + A,$$ (37)

where ζ_∞ and ζ_0 define the true and apparent positions, respectively, of a heavenly body, and A is the correction. In principle, this correction A can easily be calculated. Descartes' laws can be written, for the case of refraction in a medium of continuously varying index (or curved refraction):

$$rn \sin\zeta = R_\oplus n_0 \sin\zeta_0 = \text{constant}.$$ (38)

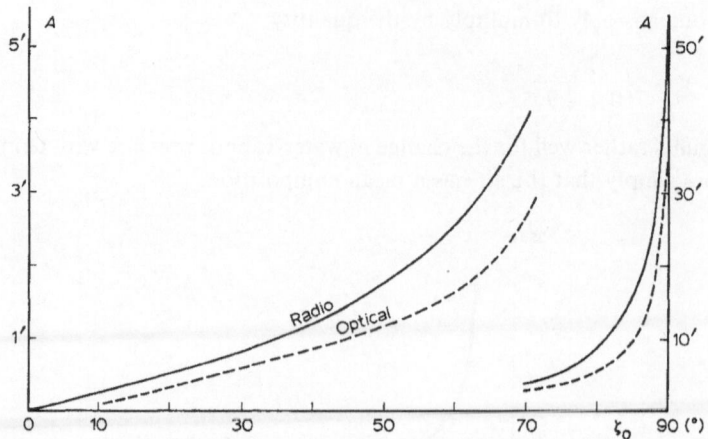

Fig. 11. Effect of refraction in the radio and optical (yellow) regions. – *Abscissa:* angle between the normal and the line of sight; *ordinate:* deflection in minutes of arc (the curves on the left go with the left-hand scale; those on the right, with the right-hand scale).

The variable ζ can be related to the variables r and θ (see Figure 12) by the formula:

$$\tan \zeta = r \frac{d\theta}{dr}. \tag{39}$$

It is important to introduce the angle V, which establishes a constant reference direction, that of the vertical at the place of observation. Clearly, we have:

$$V = \theta + \zeta. \tag{40}$$

The correction A can obviously be obtained by studying the variation of the angle V. We have:

$$A = V_\infty - V_0. \tag{41}$$

Thus we can write, using the above equations (and, in particular, taking the logarithmic derivative of Equation (38)):

$$dA = - dV = - d\theta - d\zeta = - d\zeta - \frac{dr}{r} \tan \zeta = + \frac{dn}{n} \tan \zeta. \tag{42}$$

Consequently, we can use the expression

$$A = \int dA = \int_1^{n_0} \frac{R_\oplus n_0 \sin \zeta_0}{\sqrt{n^2(r)\, r^2 - n_0^2 R_\oplus^2 \sin^2 \zeta_0}} \frac{dn(r)}{n(r)}, \tag{43}$$

where n can be calculated as a function of r by means of the above formulae for the refractive index, and a model representing the real atmosphere (that is, the variation of t, p, and p_{H_2O} as functions of r or h).

At sea level and at zenith distances of less than 75°, if the wavelength is 5800 Å, the temperature 0 °C and the pressure $p=760$ mm Hg. A can be represented by the approximate formula (due to Siedentopf and Scheffler):

$$A = (60''.4) \tan \zeta_0 - (0''.064) \tan^3 \zeta_0. \tag{44}$$

Precise calculations have also been made for $\zeta_0 > 75°$. At $\zeta_0 = 90°$, the mean refraction reaches $33'11''$ for 760 mm Hg and 10 °C. A 30° temperature variation leads to an additional variation of about 10% in this value, if the temperature is decreasing.

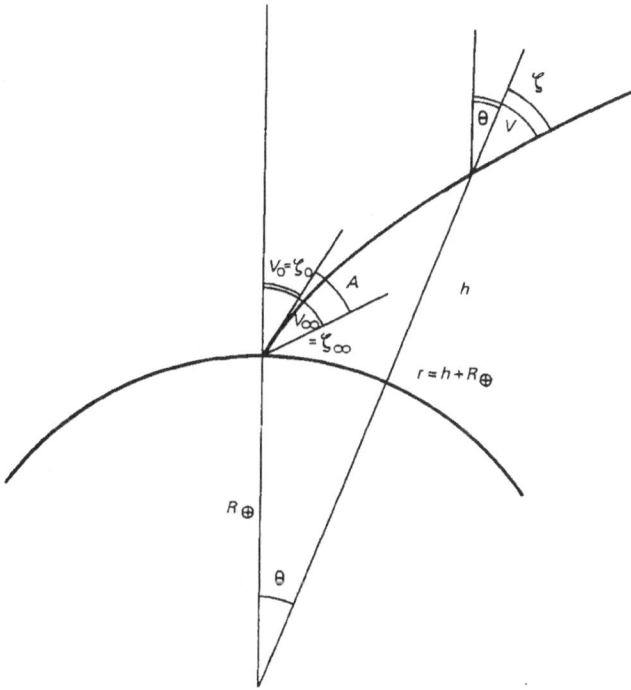

Fig. 12.   Calculation of atmospheric refraction (see text).

Astronomers have, of course, precise tables which enable them to calculate the refraction correction. This correction depends upon the spectral type of the star, for it depends on the wavelength. Unfortunately, most of these tables were calculated on the assumption that stars have a black-body spectrum; therefore they do not take into account the numerous absorption lines, discontinuities in the continuous spectrum, and molecular bands which make the energy distribution in a stellar spectrum very different from that of a black body at the same effective temperature. In practice, this produces numerous errors. Moreover, it is well known that in the vicinity of the observatory itself, the refraction undergoes random fluctuations due to variations in the pressure, the temperature, and the water-vapor content of the air – and not only near the observatory, but in the first few hundred meters of air surrounding it. Thus if the observatory is near the sea, water vapor has a greater influence on observations made in the direction of the sea than in other directions. We must remember that the accuracy required is often a thousandth of a second of arc; the permissible error at 45° from the zenith is therefore on the order of 1/50000 times the refraction correction. At the horizon, we are talking about a factor of one million!

It is true that certain observing methods reduce the problems introduced by refraction corrections. Thus astrolabes (like the Danjon prism astrolabe) make measurements only at a constant altitude, corresponding to $\zeta_0 = 30°$. And the measurements are essentially differential, referring the position of a star to those of other, supposedly well-known stars. Similarly, the Markowitz photographic zenith tube (PZT) operates only in the immediate vicinity of the zenith.

Nevertheless, certain essential problems have found no good solution. For example, the stars near the celestial equator are too low for intermediate-latitude observers in both hemispheres; thus matching up the South zone of star catalogues with the North zone is a complicated procedure, poorly carried out at the present time. Would not space research be the ideal means of eliminating such difficulties? To be sure, this is a very premature idea – what space vehicle today can determine positions to a thousandth of a second of arc? At present, spacecraft are stabilized to a much lower accuracy!

It is also important to know how the refraction changes with the altitude h. Equation (43) remains valid, but n_0 becomes the index at altitude h. We must then calculate $A(h)$, since the values calculated and tabulated above correspond to $h = 0$. If we do this, we see that the refraction is already considerably reduced at an altitude of several tens of kilometers. It is not out of the question for an observatory installed in a balloon gondola (assuming an excellent stabilization system) to make good astrometric measurements in the vicinity of the horizon.

3. The Influence of the Ionosphere

In the meter wavelength region, the ionization of the upper atmosphere has an additional effect, causing a refraction whose consequences are very important.

The index of refraction of an ionized medium is, in the first approximation, equal to

$$n = \sqrt{1 - \frac{N_e e^2}{\pi m f^2}}, \tag{45}$$

where N_e is the number of electrons (of charge e and mass m) per cubic centimeter, and f is the frequency of the waves being refracted (at large wavelengths f is often used for the frequency v, while the pulsation frequency is written $p = 2\pi f$).

In practice, this expression must be corrected if collisions have taken place among the electrons. Likewise, the presence of a magnetic field requires a modification. But since our present purpose is not the study of the ionosphere, we shall content ourselves with developing the unmodified formula as given above, from which we can at least deduce some orders of magnitude. We notice first of all that n is less than unity: $n < 1$. Next, we see that n is appreciably different from unity only if the quantity

$$1 - n^2 = \frac{N_e e^2}{\pi m f^2} = \frac{e^2}{\pi m c^2} N_e \lambda^2 \tag{46}$$

is significant.

This quantity depends on the wavelength. We have:

$$1 - n^2 \sim 0.897 \times 10^{-13} N_e \lambda^2. \tag{47}$$

For $N_e \sim 10^5$–10^6 (typical values for the ionosphere – see Figure 3), we can estimate the following upper limits:

$\lambda =$	1000 Å	1 mm	1 cm	1 dm	1 m	10 m	100 m
$1 - n^2 =$	10^{-8}	10^{-10}–10^{-9}	10^{-8}–10^{-7}	10^{-6}–10^{-5}	10^{-4}–10^{-3}	10^{-2}–10^{-1}	1–10

For wavelengths greater than a meter, the effect of refraction is thus important. The rays then propagate as indicated in Figure 13, whether they come from an astronomical source or a terrestrial transmitter. There is total reflection for an angle of incidence i such that $\sin i = n$ – that is, if

$$\sin^2 i = 1 - 0.9 \times 10^{-13} N_e \lambda^2. \tag{48}$$

For normal incidence ($i = 0$), there is total reflection only if $n = 0$ (hence for $\lambda = 100$ m, writing $N_e \lambda^2 \, 0.9 \times 10^{-13} = 1$, with $N_e = 10^5$).

This is the wavelength above which no 'ground-based radio astronomy' is possible. In reality, this limit is imposed at much shorter wavelengths, because in the F-layer N_e is much greater than the average value used in the present example.

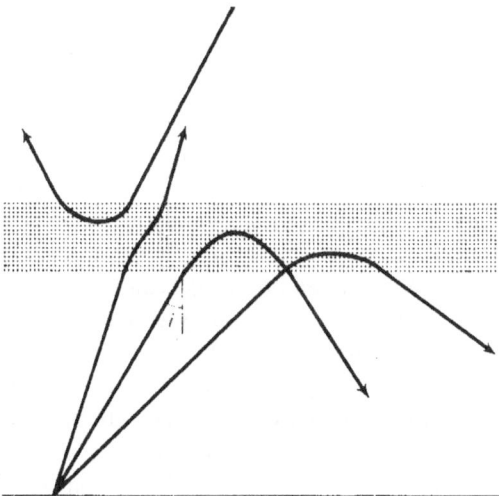

Fig. 13.   Refraction of electromagnetic waves in the ionosphere. If the angle of incidence exceeds a critical value, there is 'total reflection'.

For an angle i of about $60°$, we have $\sin^2 i = 0.5$ and a critical wavelength on the order of 70 m (with $N_e \sim 10^5$).

In fact, the measurement of the critical wavelength (or of the critical frequency) is a measure of the maximum electron density, which has been determined in this very manner.

The upper layers of the atmosphere correspond to larger values of N_e, and thus to higher critical frequencies. This fact is well represented by the classical recordings of ionospheric sounders, which show the altitude of reflection corresponding to each frequency (Figure 14).

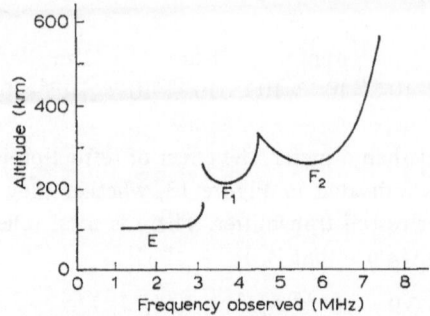

Fig. 14. Recording of an ionospheric sounding (schematic, after Mitra). – *Abscissa:* frequency observed; *ordinate:* altitude. Each curved segment corresponds to a different layer of the ionosphere.

It is therefore refraction, much more than absorption, that has a catastrophic effect on radio astronomy at wavelengths greater than about 10 m. We shall see below why this represents a real loss, and consequently why space radio astronomy should be encouraged.

4. Aleatory Refraction

Of very great importance, at visible wavelengths as well as in the radio region, are the effects of aleatory refraction, which degrades the quality of astronomical and radio-astronomical images.

We must first specify the different aspects (and their designations) of the effects due to aleatory refraction.

First of all, the stellar image – instead of being reduced to an ideal diffraction disk – is *degraded*, spread out, often forming a fuzzy spot. In addition, the image as a whole is *agitated* with an irregular motion. Finally, this effect is accompanied by *scintillation* – that is, there are perceptible variations in the intensity of the image, often accompanied by changes in the color. These effects vary according to the instruments being used, and are complicated by an integration over time when the observations are photographic. In general, they depend on the zenith distance and are much more important near the horizon.

They depend strongly upon the meteorological conditions and therefore upon the chosen site, to such an extent that a preliminary study of these effects is always made (or should be!) before choosing the location of a new observatory.

We can analyze the origin of these phenomena as follows. The inhomogeneities in the index of refraction of air have the effect that in the vicinity of the instrument, the wave fronts of stellar light are not plane but show distortions – varying with wavelength and, of course, with time (Figure 15).

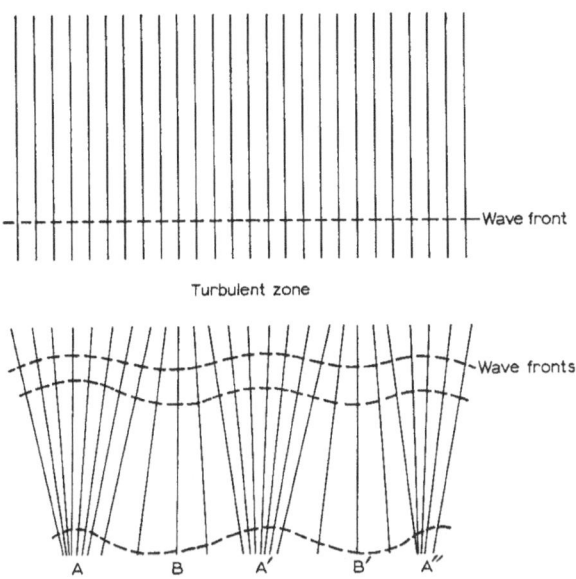

Fig. 15.   Effect of a layer perturbed by thermal inhomogeneities upon the form of the wave fronts and the illumination at the ground level.

We could imagine an experiment consisting of progressively increasing the diameter of an instrument – by means of a diaphragm, for example.

First let us see how the appearance of the image changes. If the aperture is small – smaller than the scale of the irregularities in the wave front – then the image suffers only a *degradation* and resembles the theoretical diffraction pattern. As the aperture increases, everything takes place (the wave front being warped) as if the instrument were developing astigmatism, and the diffraction rings break up little by little (Figure 16). This is what most often happens with refractors of some 20 to 70 cm diameter. If the aperture of the instrument increases further and becomes large enough to take in several irregularities of the wave front at the same time, the complicated diffraction pattern of each part of the wave front will tend to give an extended image, structureless and extremely degraded or *spread out*. This is what happens most frequently with large telescopes.

We can imagine that, for an instrument of given size, observing the appearance of the diffraction rings enables one to determine the size of the irregularities in the wave front (at least to order of magnitude).

If we follow the arrival of the wave front at the instrument as a function of time, what happens? If the instrument is smaller than the irregularities in the wave front, it is as if the wave front were plane (but not always perpendicular to the line of sight, nor of constant orientation). It follows that the image is good, quite point-like, but

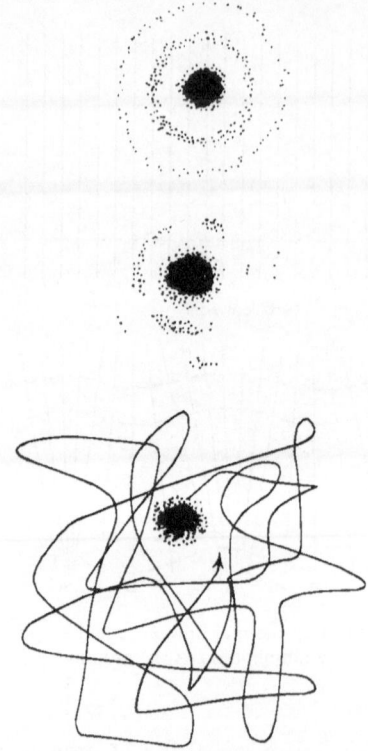

Fig. 16.   Effect of atmospheric turbulence on the displacement of images (see text).

that it moves about in a region whose apparent diameter is determined by the distortion of the wave fronts: this is image *vibration*. If the instrument is large enough to be on the same scale as the warping of the wave front, it is clear that in addition to the irregularity in brightness of the rings there will be a displacement of the center of gravity of the image. With a large instrument, the final result is equivalent to integrating the effects of a large number of plane wave fronts inclined at different angles to the direction of observation. The image is spread out over a region whose apparent diameter is $2t$, where the parameter t defines the atmospheric turbulence (or 'seeing').

In the case of photographic observations with a long exposure time, this last effect always dominates (whatever the size of the instrument).

How does the light distribution vary over the image? The distribution is far from uniform over the wave front itself (Figure 15)! For a small aperture, the image is good but its intensity fluctuates continuously: this is called *scintillation*. With a larger aperture, the light fluctuations tend to cancel out and the image becomes almost constant in brightness. To these effects, already quite troublesome for ordinary astronomical observations, there is added a *distortion* of the field. If the field is not very small, the beams coming from different stars will, at a given time, have been influenced in different ways, since they will have passed through different air masses. They will therefore give rise to images that are displaced in different ways.

It is obvious that the form of the wave front, the amplitude and scale of its irregularities, etc., depend strongly on conditions such as the thickness of the turbulent air layer, its distance from the instrument, its structure...

If we want to consider an *a priori* search for sites where the quality of the images will be degraded by the atmosphere as little as possible, we must consider the origin of the perturbations of the wave front.

First of all we note that the closer the observer is to the turbulent layer, the weaker the scintillation. At large distances, on the other hand, certain rays are convergent and a sort of focusing effect takes place. The ground is irregularly illuminated and the whole structure of irregular illumination moves over the ground, because of the relative motions of the Earth, the star, and the turbulent layer (this phenomenon is also known as 'flying shadows' and is observable during solar eclipses). Thus there is scintillation at each point. We assume that it originates several kilometers (perhaps three or four) above the ground.

But turbulent effects near the ground are caused by high-density air, and they can be large to the extent that the wave fronts will be more distorted by them than by the effects of the upper layers. They produce agitation and degradation of the images, rather than contributing to the scintillation.

The wave fronts can be distorted in a rather different way from one day to the next, and from one place to another. The value t of the seeing is, at best, $0''.1$. A value of $0''.25$ is considered very satisfactory for astronomers. But it is not unusual (on a night of high wind, for example) to find several seconds of arc. The seeing is linked with the deformations of the wave front. Its value determines the appearance of the image in a given instrument, and the time scale of motion of the image; therefore it is easy to measure with a small instrument. The 'pseudo-wavelengths' of the wave fronts can vary from tens to hundreds of centimeters; the time scale of their development is connected with their geometric scale.

The orders of magnitude can easily be calculated (using Figure 15). If the layer in question is high enough, the time scale of the fluctuations is determined by the time taken by the wave front to move relative to the instrument by a distance equal to the 'wavelength'. The structure centered on A (the instantaneous center of curvature) must replace the one centered on B. The altitude of the perturbing layer can therefore be estimated as $H = l/t$, where l is the distance AB in Figure 15.

For $l \sim 50$ cm and $t = 1'' = 5 \times 10^{-6}$ radian, we have $H = 10^7$ cm $= 100$ km. Light re-

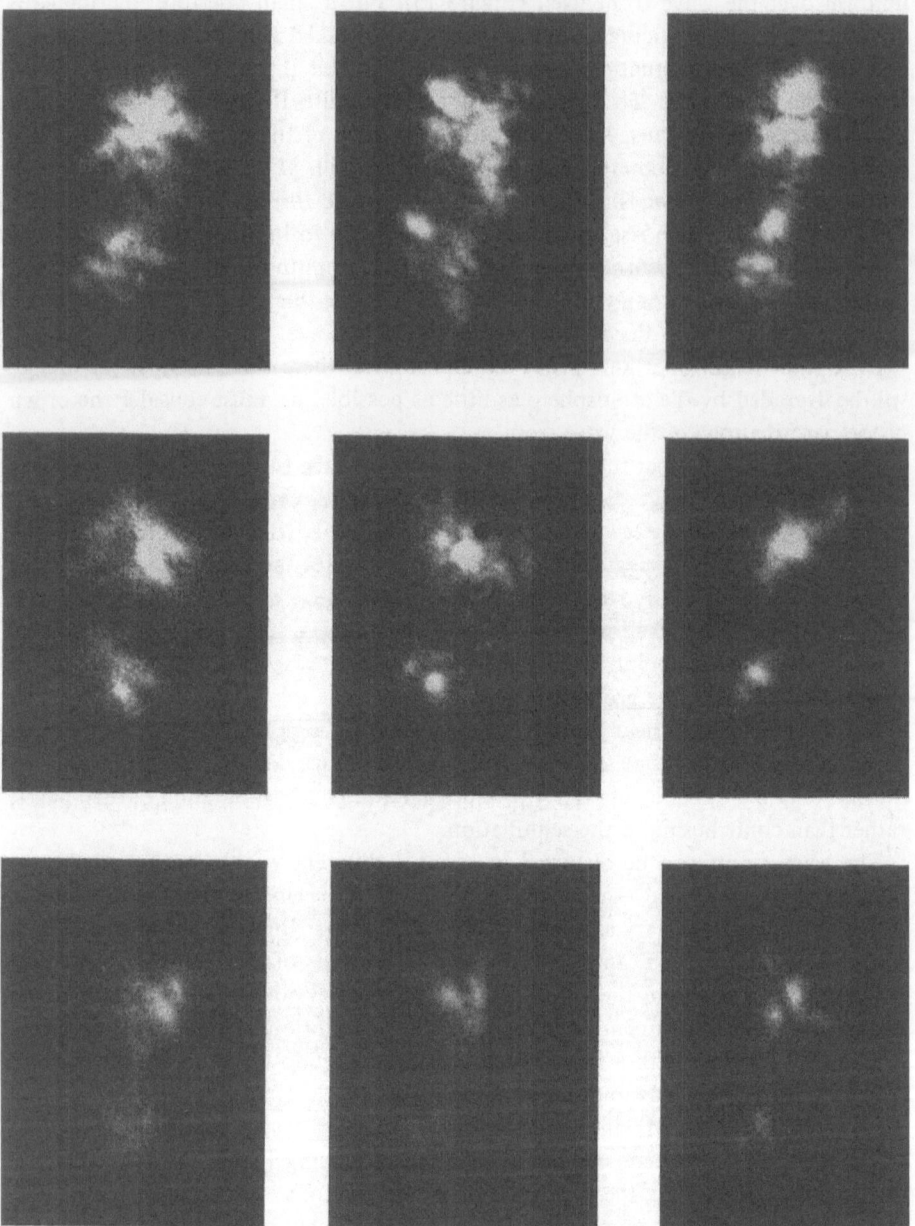

Fig. 17.   Effect of image vibration on the photograph of a double star (Pic-du-Midi). – Note that the effect is the same on both components of the star α Geminorum, of which we present nine pictures taken with the electronic camera; exposure times increase from below to above.

quires a time $\theta = H/c = 3 \times 10^{-4}$ sec to arrive from this distance. For $l \sim 10$ cm and $t = 2''$, we have $H = 10^6$ cm and $\theta = 3 \times 10^{-5}$ sec.

An exposure time longer than these values will correspond to an integration of the image. A shorter exposure time (the electronic camera now makes this conceivable) will enable us to obtain an image that is displaced, but smaller. The displacement will be the same for two very similar directions. Thus we will have (Figure 17) a good method of measuring the separation of double stars, despite the atmospheric turbulence. Similarly, the solar granulation can be photographed well, even from the ground, by using sufficiently short exposure times and selecting out the good images – despite their agitation. (See Figure 17.)

Figures 18 and 19 show the spatial scale and the time scale of the scintillation. Figure 18 presents, after Ellison and Seddon, the variation of the intensity dispersion $y = \Delta I_D^2 / \Delta I_0^2$ with the diameter of the instrument, where I_D is the intensity corresponding to an instrument of diameter D, and I_0 the intensity corresponding to a 'limiting instrument' of zero diameter. We see that the order of magnitude is a few centimeters. Figure 19 presents, after Protheroe, the frequency spectrum of the

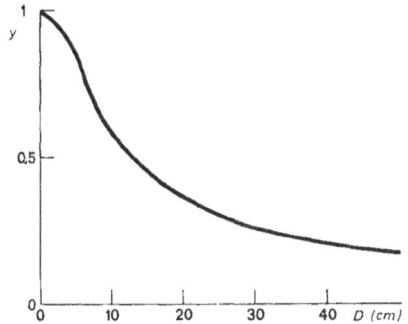

Fig. 18.   Spatial scale of atmospheric scintillation (after Ellison and Seddon).

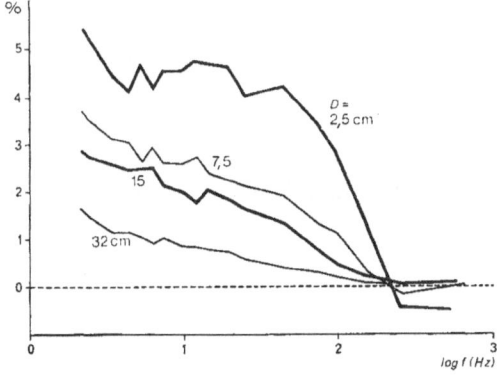

Fig. 19.   Time scale of atmospheric scintillation (after Protheroe). Results of experiments made with different instrumental apertures.

scintillation for different instrumental apertures; here the order of magnitude is a few thousandths of a second.

If we are dealing with a structure near the ground, the motion of the wave front with respect to the instrument is connected with the velocity (wind) of the perturbed layer, which may be on the order of a few kilometers an hour. If l is the scale of the irregularities in the wave front, we can define a 'characteristic time' of $\tau = l/v$; its order of magnitude is 4 hundredths of a second. This is the time scale for vibration, which is generally slower than scintillation.

A study of the properties of the image thus enables us to determine more or less precisely the altitude of the layers responsible for them. Now, we would like to eliminate all these deleterious effects of the atmosphere: (a) scintillation disturbs precise photometry, and integration in time is a solution only for the brightest stars – for faint objects, the sky background limits the potential of this technique; (b) chromatic scintillation disturbs spectrometry in a similar fashion, and again there is a limit to what can be accomplished by integrating in time; (c) the vibration and enlargement of the images perturb determinations of absolute positions, and thus limit the precision of star catalogues; (d) the degradation of the images disturbs the detection of faint companions, as well as causing other serious difficulties.

It is obvious that the turbulent layers near the dome (at its very entrance) and even inside the dome) – and inside the instrument itself – can be eliminated, or at least minimized, by well-designed buildings. The low-altitude turbulent layers are mainly due to temperature effects, which are present when the ground is significantly warmer than the air above it. These effects can be minimized by the choice of appropriate vegetation around the observatory. But the medium- and high-altitude winds, responsible for large turbulent effects, cannot be eliminated. In particular the *jet stream*, located at an altitude of several kilometers, often causes turbulent zones.

For this reason, the complete elimination of the effects of atmospheric turbulence on astronomical observations can be obtained only at altitudes greater than about ten kilometers. We shall return to this point.

As for radio waves in the centimeter and millimeter wavelength regions, scintillation is also very troublesome. (The vibration of the images is not noticeable because the methods used to determine position are still not precise enough.) This scintillation is principally due to the layers of the ionosphere located at altitudes between 300 and 500 km. The displacement of the images is on the order of several minutes of arc; the time scale for intensity variations is on the order of several minutes of time. The scale of the irregularities in the wave front near the ground is on the order of a kilometer.

CHAPTER V

THE PARTICLE BARRIER

The Earth's atmosphere does not resist only the passage of electromagnetic radiation – or, at least, the passage of this radiation in undisturbed form. It also resists, through its density and especially through its electromagnetic properties, the more or less energetic particles that approach the Earth from space, and principally from the Sun. In particular, it resists the passage of the so-called 'primary cosmic rays' – consisting of charged or neutral particles – and, in a quite different way, the passage of meteorites (see Chapter VI).

There is no doubt that most of the high-energy charged particles come from the Sun, and particularly from the most active regions of the Sun. At the Earth's distance from the Sun, the number of protons that can be received per square centimeter per second depends on the particle energy; and this energy distribution is determined by the origin of the protons. Figure 20 (which has been drawn up chiefly from results obtained by space research) shows the magnitude of the relevant energies in different cases. Note the preponderance of the solar particle flux in comparison with the non-solar cosmic flux. Note also the range in particle flux arriving at the Earth, from cosmic rays originating in the Sun but not in chromospheric flares to the high-energy jets associated with the brightest flares. The jets are distinguished by their density as well as by the energy distribution of the particles they contain. In addition to the particles that make up the solar wind and are associated with active phenomena, we

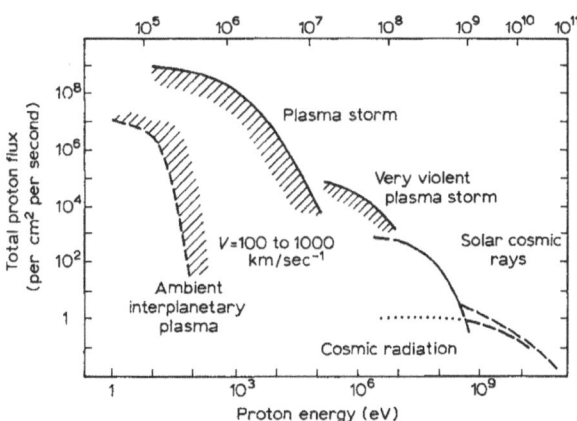

Fig. 20.   Energy spectrum of solar particles in the vicinity of the terrestrial orbit (schematic, after Obayashi).

also note the existence of an ambient plasma in the interplanetary medium, rich in very low-energy particles. In sum, we are dealing with an 'average' medium upon which are superimposed, from time to time, jets of active particles.

These data pertain to the protons (hydrogen nuclei). But heavier nuclei are also contained in the particle jets, whose composition has been found to be similar to that of the Sun. Sounding rockets have also made it possible to measure the electron flux. But it has become clear – and this is quite a remarkable phenomenon – that the proportion of heavy nuclei appears to decrease as the energy of the particles increases.

How does the Earth react to these charged particles of various energies?

It essentially prevents us from observing them, and forms an almost impenetrable barrier to all except the highest-energy particles, by virtue of its magnetic field. Research on this subject has been going on for a long time. As soon as scientists began to study the terrestrial magnetic field, they realized that this field – however weak it might be – would prevent charged particles from traveling in a straight line. Simple models of the terrestrial magnetic field were constructed as early as the end of the last century; at that time it was a question of explaining the polar aurorae, terrestrial effects of the arrival in the upper atmosphere of charged particles from the Sun. Birkeland (1896) constructed an experimental model of the Earth, the 'Terrella'; shortly afterwards (around 1910), Störmer attempted a detailed theory of the motion of charged particles in the terrestrial field. The problem has, of course, been developed considerably since then. Störmer showed how particles could be captured by the terrestrial magnetic field without being able either to escape or to reach the surface of the Earth. We now know that the zones discovered by Van Allen in 1958, and bearing his name, are just such regions where the solar particles are 'trapped'. These are medium-energy particles. Particles of higher energy penetrate further and escape this trap – only to fall into another, that of collisions.

For upon encountering the particles of the lower atmosphere, the electrons and primary cosmic radiation induce ionizations, disintegrations, and a whole range of phenomena such that an energetic particle gives rise to a *shower* of particles (including photons). This shower of particles gives rise to secondary showers, and so forth. At ground level, only an infinitesimal proportion of the primary radiation is observed. Only very high-energy mesons are directly observable – those with energies in excess of 10^{10} eV!

It is impossible to present in this chapter a detailed study of the interactions between the Earth with its atmosphere, and the particles of all kinds that surround it. But we are obliged to acknowledge this interaction, which affects particles with energies up to several billion electron volts. This amounts to saying that any study of the local medium surrounding the Earth must involve space research. Figure 21 shows how we now imagine the structure of the magnetic field around the Earth, perturbed by the solar wind. Only space research has made it possible to obtain such results. Need we say more?

As for the energetic neutral particles, the neutrons and neutrinos of the cosmic radiation, their penetration is doubtless not affected by the terrestrial magnetic field.

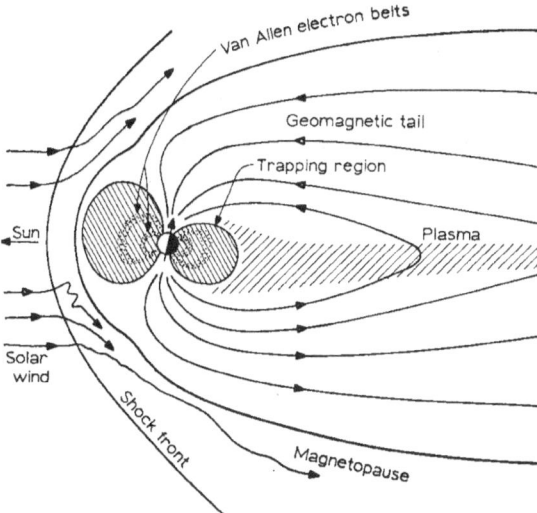

Fig. 21. Interaction of the solar wind with the geomagnetic field (north-south cross section of the magnetosphere).

The most energetic neutrons are observed at ground level; the neutrinos are so penetrating that it is virtually impossible to observe them. No doubt the first progress in this field will be the result of improved detection methods. But this question is obviously of such great importance (because of the study of nuclear phenomena in stars, which produce neutrons or neutrinos that can travel from the center of the star to its surface without appreciable perturbation!) that every effort should be made to improve the detection techniques, on the ground and in space, in order to advance the astronomy of neutral particles.

CHAPTER VI

THE FUSION OF METEORITES

Observations – throughout the centuries – of shooting stars, and the discovery of 'aeroliths', or stones fallen from the sky, imply that the Earth passes through regions of space rich in small solid bodies, traveling in the interplanetary medium. The present state of our knowledge can be summarized by saying that meteorite swarms most often originate in former comets. Thus Biéla's comet divided in two, in the course of its successive passages near the Earth, and later broke up into a collection of small bodies (Figure 22). The velocity dispersion at the time of the probable explosion, taking place at a given point in the trajectory, produces the progressive spreading out of the debris, called meteorites, over the whole of the former cometary orbit. Thus every year at the same time, when the Earth crosses the orbit of a comet that has disappeared, it actually crosses a more or less dense swarm of meteorites.

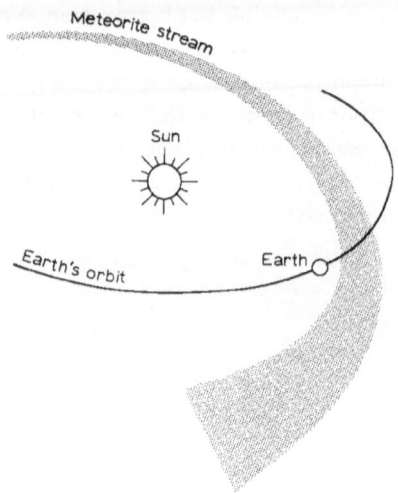

Fig. 22. Meteorite stream.

There are the Bielids, named after the comet whose history is related above. There are the Leonids, the Geminids, and many others, named simply for the region of the sky that corresponds to the orientation of their orbit in the neighborhood of the Earth – that is, the region of the sky in which the corresponding 'shooting stars' seem to originate...

Visual observations of shooting stars give us information concerning the meteoritic flux. More recently, studies (by radar) of the trails of ionized gas produced in the upper atmosphere by meteorites have made it possible to measure objects of smaller mass than those associated with the optically observed shooting stars. But it is quite clear that the interplanetary medium is full of dust, which is vaporized upon arriving in the upper atmosphere. The smallest dust particles escape the observation of the radar specialist as well as of the optical astronomer; and no doubt they constitute the major part of the mass of dust surrounding the Sun, which is made up not only of the debris of comets and asteroids but also of the ancient, primordial dust whose existence probably preceded the existence of the Earth itself – that dust which formed the 'primeval nebula' and which doubtless still exists in enormous quantities in the interstellar medium, where stars are born.

How then can we obtain a better knowledge of the interplanetary dust, the meteoritic density in the interplanetary medium, and the micrometeoritic density? The question is all the more important now that the lives of our astronauts are at stake. A rather heavy meteorite, a bombardment by small meteorites, or a sweep of thick dust could endanger their equipment and even their existence.

Certain fine dust particles fall without being destroyed, and their accumulation at the bottom of the oceans is a measure of the meteoritic flux. But most particles can be observed only from rockets and satellites. Figure 23 gives a summary of our

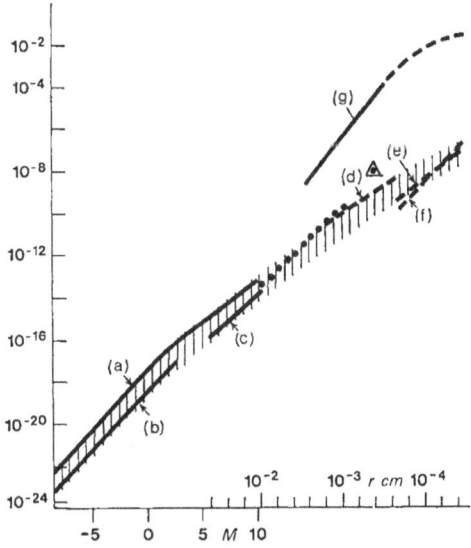

Fig. 23.   Distribution of meteorite sizes (after T. Kaiser). – *Ordinate:* number of particles per m³, of radius greater than *r* cm (*abscissa* at right) or, for large meteorites, capable of producing a shooting star of magnitude greater than *M* (*abscissa* at left). Curves (a), (b), and (c), derived from various authors, correspond to sporadic meteors. Curves (d), (e), and (f) represent the zodiacal light. The points correspond to the accretion of cosmic grains. Curve (g) represents observations by satellites; the extrapolation of the satellite observations to the distance of the lunar orbit produces the triangle.

present knowledge. It shows the distribution of meteorites as a function of their size. Data concerning their velocities would naturally be an important addition to our knowledge; but this varies a great deal from case to case, no doubt determining the rapidity of the destruction of a given object, and the magnitude of its terrestrial effects. The mean energy of the interplanetary particles is far from well known; much remains to be done in this field through research with space probes.

PART II

A PROVISIONAL AND PARTIAL INVENTORY OF SOME OF THE INFORMATION TO BE ACQUIRED BY SPACE RESEARCH

Une pierre	One stone
deux maisons	two houses
trois ruines	three ruins
quatre fossoyeurs	four grave-diggers
un jardin	a garden
des fleurs	some flowers
un raton laveur	a raccoon
une douzaine d'huîtres un citron un pain	a dozen oysters a lemon a roll
un rayon de soleil	a sunbeam
une lame de fond	a ground-swell
six musiciens	six musicians
une porte avec son paillasson	a door with its mat
un monsieur décoré de la légion d'honneur	a gentleman wearing the Legion of Honor
un autre raton laveur	another raccoon
un sculpteur qui sculpte des napoléon	a sculptor sculpting napoleons
la fleur qu'on appelle souci	the flower called marigold
deux amoureux sur un grand lit	two lovers on a double bed
. . . .	. . . .
et...	and...
plusieurs ratons laveurs.	several raccoons.

JACQUES PRÉVERT

Paroles

The reader may be wondering what the bizarre inventory of Jacques Prévert has to do with the business at hand... We have seen, at great length, how the atmosphere is an impenetrable wall for astronomers – opaque, distorting, perturbing – and how the astronomical observatory is in reality a prison, where only a little dormer-window open onto the sky, up above the roof-tops...

To be sure, the intelligent imagination of the imprisoned astronomer, aided by an inspired glance at the sky we see, has amassed an enormous fund of knowledge since the time of Hipparchus, Tycho, Newton, Herschel, and many others.

What more, then, can be added in the various fields of astronomy by an escape into space?

Before discussing actual achievements, we must undertake a *prospective* examination (the French word 'prospectif' is used by the author here, although he deplores the use of this fashionable neologism in contexts of all kinds. Actually, the English word 'prospective' – in the sense of anticipating the future – is derived from Latin through a long-obsolete French form; the French rejuvenation of the word is obviously patterned after the English).

Our prospective examination should be systematic, and proceed according to the rules so well defined by Zwicky in his *Morphological Astronomy*. In this sense, it should be an *inventory*. I shall confine myself to giving a few examples, more or less following the outline of Part I.

CHAPTER I

BEYOND DIFFUSION

1. The Solar Corona, Observed Without the Handicap
of a Diffusing Atmosphere

We know that the solar corona was discovered during eclipses of the Sun. At such times, the luminosity of the sky is about 7 to 8 magnitudes fainter (that is, a factor of around 10^3 in intensity) than in broad daylight.

The visible light of the corona is known to be composed of several distinct types of radiation: first of all, the 'white' K corona, which corresponds to scattering of the solar light by the electrons that constitute the major part (in number of particles) of the corona. Then, the 'white' F corona, which is associated with scattering by the dust of the outer corona. And finally, the monochromatic emission lines, which are produced by the very highly ionized elements of the corona and form an important part of its visible spectrum.

Figure 24 shows the intensity of the white-light radiation as a function of the logarithm of distance from the Sun, extrapolated outwards with data obtained from measurements of the zodiacal light. Of course, this figure is only schematic, for the simple reason that we have presented only average values, while the corona is a highly inhomogeneous medium!

We see that in a balloon, we can obtain results comparable to those obtained during an eclipse. In a satellite, there would be practically no limitation – the zodiacal

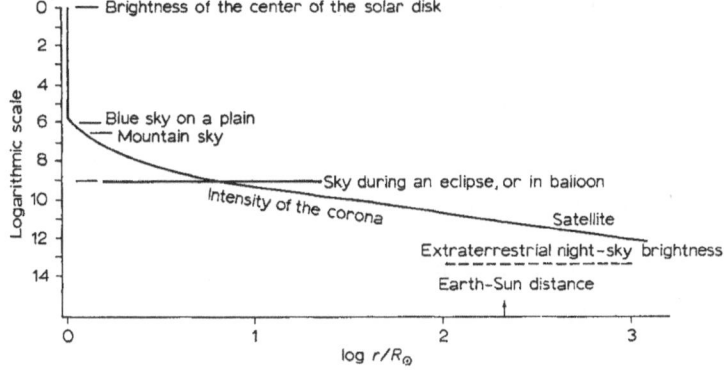

Fig. 24.   Distribution of intensity around the Sun (scales are logarithmic).

light itself would present the chief obstacle to observation of the distant sky, at least in the plane of the ecliptic, where Figure 24 applies: for we recall that the corona, like the solar system itself, is highly flattened.

In the same spirit, we can say that the apparent surroundings of the Sun will be easier to observe: the relativistic distortions of the star field behind the Sun will be easier to measure. From these, we can expect a confirmation of Einstein's predictions concerning the deflection of light rays by the mass of the Sun. Improvements in the theory of relativity may result from these measurements, whose success is assured once the atmospheric diffusion is eliminated.

2. Distant Galaxies and Faint Objects

Still more important are the gains to be made in studying the most distant – and therefore the faintest – sources: those galaxies and other objects that populate the outer reaches of the observable universe.

We know that the universe is expanding, according to the spectrographic measurements accumulated since 1920 by Hubble and his associates: the galaxies are moving away with a velocity of recession proportional to their distance. But the interpretation of Hubble's measurements has been disputed – thus it appears necessary to find other methods of verifying the expansion. Hubble himself thought of counting distant galaxies: at distances of billions of light years, they represent their corner of the universe as it was billions of years ago. Consequently, if there is expansion (continuous expansion, on this time scale), the galaxies observed should be less numerous in the vicinity of our own Galaxy than they seem to be in a similar volume at great distances.

Unfortunately, we are greatly restricted by the sensitivity of our equipment, and even more so by the brightness of the night sky, which was the subject of some earlier remarks (Part I, Chapter III).

The relationship between magnitude and number of distant galaxies can in principle be studied with the data acquired from ground-based instruments. But we must realize that such counts are very difficult and, consequently, to be treated with great caution as regards the faintest galaxies. The principal reason for this is the existence of great fluctuations in the space density of the galaxies: there are clusters of galaxies, and probably clusters of clusters.

Another difficulty results from the fact that, since galaxies have real diameters differing greatly from one another, the limiting magnitude depends on the type of galaxy studied; moreover, the distribution of the number of galaxies as a function of magnitude depends both on a distance distribution and a size distribution – it is very difficult to separate the two effects.

Finally, the distant galaxies may even be, statistically speaking, in a more advanced evolutionary state than the galaxies that surround us – is the comparison then significant?

If all the galaxies were of the same intrinsic brightness E_0, their brightness E would

lt to calculate. But
ıplicit in the follow-
tions. For our part,
ın a general idea.
₀ = r/c to take into
ace of observation.

ice r_0 by the formulae:

$$/E \tag{1}$$

$$\tag{2}$$

(8)

he density at each point (in number
ɛr of galaxies brighter than a certain
$ = (4\pi/3)r_0^3$ – is:

tant, which appears
ears and the unit of
ɔlementary term ΔN
from unity – that is,
ɔonds to about two
.e chance of showing
ɛ – in the framework
ɛsolutely precise and

$$\tag{3}$$

urse not valid. We must take into
ion of time, $n(t)$; let t_0 be the actual
phere of radius r_0. But it no longer
ı magnitude m: they have run away.
ince r from the point of observation,
on of the universe. The rate of ex-
ɔalaxy is, according to Hubble,

$$\tag{4}$$

ies of any given type (in particular,
ˆ we assume they all have the same

$$\tag{5}$$

ı 'apparent' distance r: the galaxies
ˊ, and appear to be located at this
ɔt instantaneous. But Hubble's law
ɛ Hubble constant refers to such

ın, the number of galaxies (of any
ı m, which was used to define the

$$\tag{6}$$

$log_{10}\alpha$ 4

y). – D: observed distribu-
fined in the text.)

ʻrite:

$$lr = N_{\text{static}} + \Delta N. \tag{7}$$

The quantity ΔN (negative, if there is expansion) is not diffic
we shall give no more quantitative information about it than is i
ing graphs, leaving the reader to amuse himself with the calcul
we believe them to rest upon data too imprecise to give more th

We shall merely point out that one can obviously write $t-$
account the propagation time from the point observed to the
We find:

$$\Delta N = n_0 f(r_0) = n_0 F(m - m_0).$$

The calculation can be carried out if we know the Hubble co
in the expression for the function F. If the unit of time is 10^9
distance is the megaparsec (10^6 pc), we have $H \sim 0.2$. The con
is significant only if the exponential in Equation (5) is differen
roughly, for $3(H/c)r \geqslant 1$, or $r \geqslant 500$ mpc. This distance corre
billion light years: at lesser distances, the counts have rather lit
the slightest expansion effect. At greater distances, there is hop
of a theory which, unlike the one we have just sketched, is

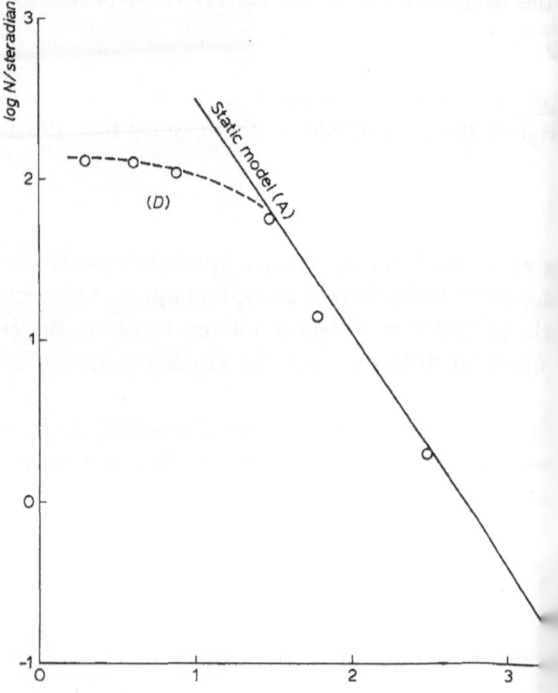

Fig. 25.   Space distribution of galaxies (cumulative diagram, after Zwic
tion; A: static model (no expansion). (*Abscissas* and *ordinates* c

the magnitude (in the preceding examples, only one quantity could be determined for a distant object: apparent diameter or apparent luminosity – not both at once!). From this it is inferred that the dispersion is intrinsic. It is possible to construct a distribution curve of absolute magnitudes (Figure 28) which certainly explains much if not all of Figure 27, and which has nothing whatsoever to do with the expansion of the universe! We see how useful it would be to gain even one magnitude, for an extension of counts of this kind.

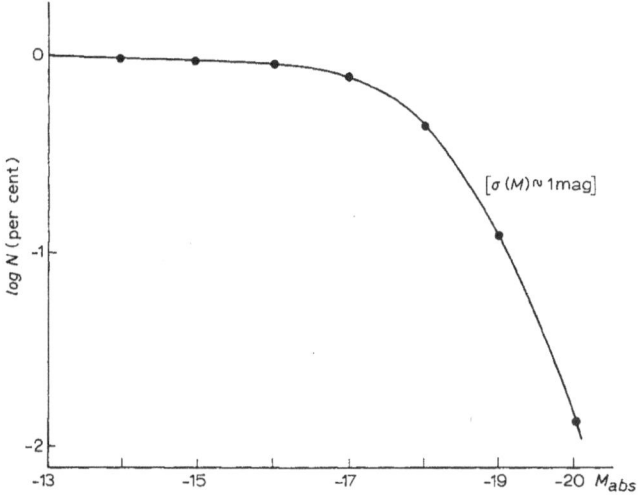

Fig. 28.   Distribution of absolute magnitudes of galaxies.

Another important justification for seeking such a gain is the need for a precise determination of the velocity-distance relation – that is, the extension of Hubble's law. Thus Figure 29 presents the measurements as a function of magnitude. It concerns 18 clusters of galaxies, and the magnitude indicated is that of the brightest galaxy in the cluster. We see that the ambiguity between different cosmological models can be resolved only by an improvement in the limiting magnitude. What improvement is possible in these different types of statistical studies?

In fact, since galaxies have a perceptible apparent diameter when observed with powerful instruments, one can gain at best about a magnitude (see Part I, Chapter III, Section 4) in the photovisual region. More can be gained, in principle, in the radio region. But there will be no gain in resolving power, for the objects to be studied have an apparent diameter larger than the diffraction disk. Resolution can be gained only when the instrument is used to study galaxies whose real apparent diameter is less than that of the diffraction disk of the radio telescope. It is open to discussion whether or not this is possible for a given instrument, considering the restrictions imposed by the sky brightness, even using a satellite or a lunar station. A detailed prospective study of this problem remains to be made, at the dawn of the conquest of space.

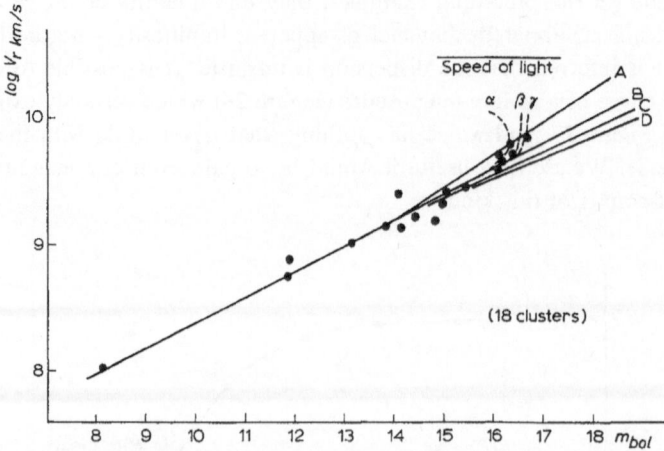

Fig. 29. Hubble's law from clusters of galaxies (after Neymann and Scott). The letters A, B, C, D correspond to various cosmological theories which do not take into account any selection effect. Model D is the stationary model; the others are expanding models, with A having the smallest time constant. The letters α, β, γ correspond to theory B and to various observational selection effects. These effects are defined by two magnitudes, m_1 and m_2, and a number n. In order for the cluster to be rich enough to be considered a cluster, it must contain n galaxies brighter than m_1; on the other hand, the redshift is measurable only for the brightest of these n galaxies – that is, for those whose magnitude is less than m_2 (where $m_2 < m_1$). In all three cases, $m_1 = 21.5$ and $n = 25$. For α, $m_2 = 19.5$; for β, $m_2 = 20.5$; for γ, $m_2 = 21.5$. For curve B, there is no selection effect.

3. Other Faint Objects

If galaxies, for which the limiting magnitude is around 24.5 (as opposed to 23.5 with the largest existing telescopes, for point-like objects) are certainly one of the most fruitful fields for space exploration, although only one magnitude is to be gained, the ideal subject is the study of point-like objects and near-by objects.

What objects fall into these categories?

(a) *Planets of other solar systems.* – If Jupiter revolved about a star located ten parsecs away, it would have an apparent visual magnitude of about 28. Even if space research made it possible to separate the planet from the star (by eliminating the problem of diffraction, with some kind of giant instrument on the Moon), this brightness would probably still be too small to observe directly. But we shall return to the potential 'separating' power offered by space research.

(b) *Asteroids in our solar system.* – In the asteroid belt that lies, to a first approximation, between the orbits of Mars and Jupiter, there are about 200 known asteroids greater than 100 km in diameter, and 500 between 50 and 100 km in diameter. The total number of known asteroids can be estimated at 1700 (as an order of magnitude). This number is limited only by the limiting magnitude defined in Part I, Chapter III, Section 5.

There is associated with this limit a certain mean diameter of the asteroid. Gaining a factor of 2 in diameter would be equivalent to gaining 1.5 magnitude; this corresponds to gaining a factor of 2 in the number of known asteroids, according to a simple extrapolation of the statistics we have just introduced. To gain a factor of 10 in the number, a gain of about 5 magnitudes would be required. We see that space research would make it possible to multiply the number of known asteroids by a factor of this order of magnitude – even if it were only space research carried out at the edges of the Earth's atmosphere. Probes (of the Mariner type) that go beyond the orbit of Mars could obviously do much better.

One might naturally wonder whether such a gain would have any real astronomical value. The study of asteroids often seems to be one of the most petrified branches of astronomy, for asteroids have been very advantageously (they say) replaced by artificial satellites in many of their astronomical applications. But this is to value too lightly the great cosmogonical interest of the problem. A detailed study of families of asteroids that are contemporary with each other (the Hirayama families) enables us to trace certain explosions in the past. These studies are essential, and are often lacking in data.

(c) Next will come the study of *space vehicles* (probes and satellites) themselves: we will be able to follow them much farther than is possible today.

(d) Finally, the study of *cold stars*, of low intrinsic brightness, will be greatly extended. These stars are of very special interest, for they are objects at the beginning of their evolution (red dwarfs) or the end of their evolution (white dwarfs, neutron stars and pulsars). It is difficult to predict how many of our evolutionary theories may be overturned by an improved knowledge of these objects. It is perhaps in this field that the most important research on faint objects will be done, bringing new elements to our ideas about the extreme stages of evolution. These are the most important studies in the fields of classical optical astronomy and classical radio astronomy that space research will further by freeing telescopes from atmospheric diffusion.

But we shall see that our pretensions are even greater, thanks to possible extensions to the forbidden wavelengths of the ultraviolet, the infrared, and the meter and kilometer waves (see Chapters III, IV, V, and VI below).

CHAPTER II

BEYOND TURBULENCE

We have seen that one of the most important limitations of astronomy is imposed by the spreading of images due to atmospheric turbulence (aleatory refraction), and by the deflections due to astronomical refraction.

To get a specific idea, we shall suppose that the seeing is on the order of 1″ in the following examples.

It is quite clear that in a number of fields, the elimination of this restriction would be of basic importance. For then, we recall, the spreading of the image would be due only to diffraction, with a radius of $1.22\lambda/D$. To minimize this value, we would have to work with large objectives (big D), at wavelengths as short as possible (small λ). But note that balloons, which carry rather large instruments, would do the trick in a number of cases.

1. The Surface of the Sun

The problem of the solar surface is the first subject to be approached with success, even at this early stage, using space observations from balloons. We know that the surface of the Sun is far from homogeneous. The solar granulation, observable in the visible region, is the photospheric manifestation of a deeper convective zone, in which are formed the waves that propagate towards the exterior. Observations of the granulation are fundamental for anyone who wants to understand the mechanism of energy transport in the solar atmosphere. Because of its nearness, the Sun is the only star in which one can observe this phenomenon – the importance of these observations is all the greater for this reason, despite the theoretical difficulty that will probably occur in generalizing the results.

Now the scale of the known inhomogeneities is on the order of 0″.5, and it is precisely images of this size that are most subject to agitation, even more vexing by day (under the influence of the Sun's heat) than by night. Thus the images of the granulation are scrambled, as it were, and the measurements we want to make (intensity of the radiation, radial velocities, profiles of spectral lines, even magnetic fields...) in the granules and between the granules, are completely washed out – all the more so for experiments requiring a longer exposure time. Consequently it is impossible to go from the measurements to the desired quantities – since the per-turbing function is unknown and variable, the deconvolution process encounters insuperable difficulties. Only a gain in resolution would permit physically meaningful determinations, which would then be directly usable in understanding the Sun.

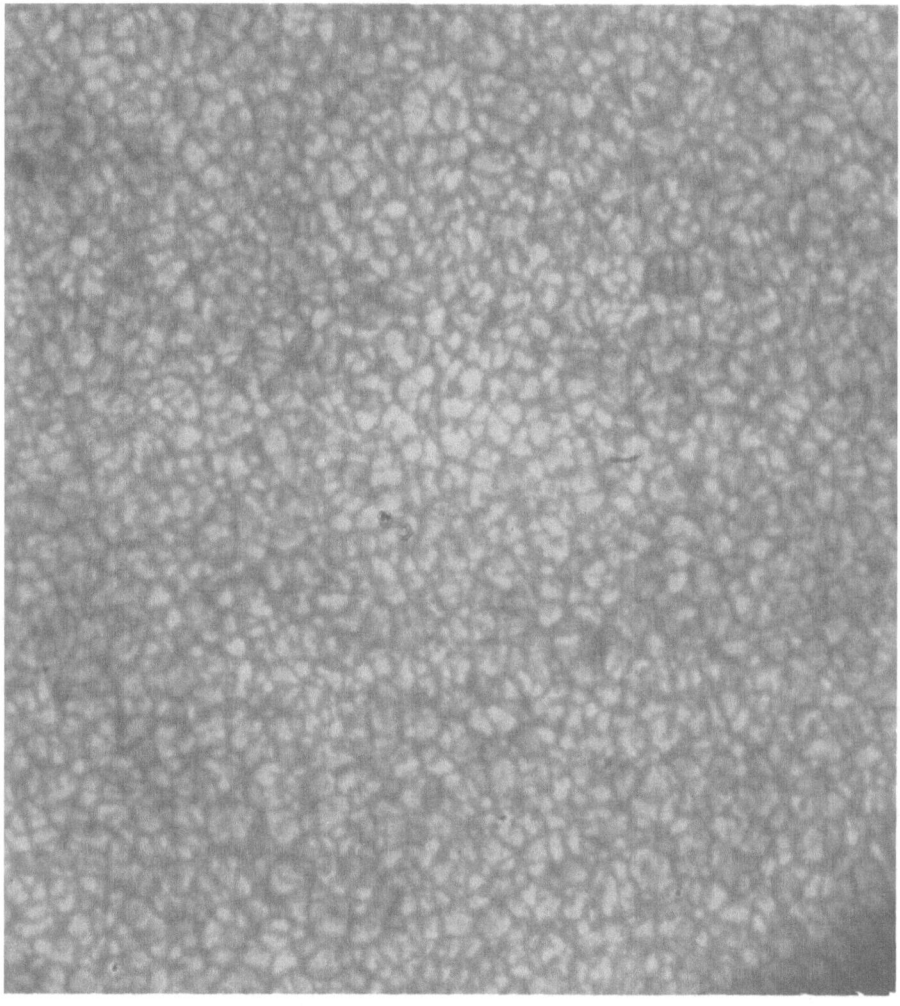

0"   10"

Fig. 30. Solar granulation, photographed from a balloon by Schwarzschild (Project Stratoscope of Princeton University, sponsored by NSF, NASA, and ONR). – The scale (in seconds of arc) is given below right. A photograph with such good definition is only rarely obtained from the ground, because of image vibration.

To be sure, it is already possible to obtain such measurements in the high mountains (at Pic-du-Midi, for example). But a balloon experiment, though more difficult
to get off the ground, is nonetheless a better research tool, even with a telescope of
rather small aperture. It would be a fine idea to exploit these possibilities to the fullest.
However, the only successful experiments in this field to date have been very limited.
Figure 30 shows an example of the measurements made by Schwarzschild and his
co-workers. There is no doubt that a complete program of this kind (combining the
ability of the big, modern balloons to go beyond the turbulent zone of the jet stream
at 15–16 km, with the ability of the electronic camera to reduce the exposure time
to an absolute minimum) could give precise information on all aspects of the inhomogeneous structures of the solar photosphere – and even those of the chromosphere
and the corona, by operating in the infrared or, better still, in the ultraviolet.

2. The Solar System

It is easier, as we all know, to go and observe Mars directly (and perhaps even Venus!)
than to put into Earth orbit a space observatory stable enough, and with a big enough
telescope, to give an appreciable improvement in resolution over terrestrial observations. But on the other hand, it is impossible to imagine such an expedition to the
outer planets, to their satellites, or to the asteroids. Yet, improving our knowledge
of the surface of planets like Neptune or Pluto, or asteroids like Ceres and Pallas,
would perhaps give important information on the origin of the solar system, insofar
as we would be better informed on the origin and evolution of these bodies themselves.

3. The Planets of Other Systems

From space, there is obviously a certain possibility of observing such bodies, even
in distant systems. Since the sensitivity of the observations would be increased, the
problem would not be so difficult as it is on the Earth. Remember that the separation
would be several seconds of arc – the problem would be to dim the brightness of the
central star sufficiently. It is a difficult problem, but does not require any great
prowess in the matter of spatial resolution, provided the planetary companion is
bright enough (definitely brighter than Jupiter!).

4. Stellar Diameters

At the present time, interferometric methods have given us only a very small number
of stellar diameters – again the atmospheric turbulence gets in the way. In the most
favorable cases (like Betelgeuse), the apparent diameter is on the order of 0".05. The
resolving power of the instruments is not the limiting factor – even on the Earth! But
to eliminate seeing, one has to resort to very sensitive photon-counting techniques
with elaborately designed receivers. At an extra-terrestrial observatory with, for
example, a one-meter telescope, the smallest measurable diameter would be on the

order of twice the diameter of the diffraction disk, or:

$$\varepsilon \sim 5\lambda \times 10^{-5}.$$

If $\lambda \sim 3000$ Å, ε is on the order of 1.5×10^{-6} radian, or $0''.5$ (after all, why not work in the ultraviolet so as to reduce diffraction effects?). The image would then be perfect, but the diameter would still not be measurable. If one were to use two tele-scopes 100 m apart, the smallest measurable diameter would be $0''.005$. And since there would be no problem in maintaining the coherence of the two beams, and no atmospheric turbulence to require elaborate instrumentation, we may assume (even though the research would certainly be difficult, and costly!) that the high probability of success and the relative simplicity of the operation might well make it very profitable.

For the measurement of stellar diameters, we can also consider another experiment that has been tried on the Earth, but found to be too difficult because of the atmo-spheric turbulence – the photoelectric measurement of the occultation of a star by the sharp, dark limb of the Moon. When an occultation occurs, the diffraction patterns are perturbed by the existence of the finite apparent diameter of the star. To derive the diameter from these patterns is a rather subtle exercise in numerical deconvolution – but it is possible, at least in principle.

These experiments can, of course, be performed with photoelectric receivers, and the results telemetered back to the Earth.

5. Double Stars

It is well known that the measurement of visual double stars is our best method (and the only really correct one) of determining stellar masses. Now the number of these objects is very large, but the number of usable pairs is quite limited – on the one hand because of the limiting magnitude of the instruments, and on the other because of the small apparent separation of the components. Using an extra-terrestrial obser-vatory, one can gain on both counts. A rapid theoretical calculation shows that it is by no means absurd to suppose that the number of usable binaries could be multi-plied by a factor of 100 or 1000.

The difficulty arises in that observations of this type, at present entirely visual (if we except the recent, and promising, attempts to make use of the electronic camera), are very difficult to render completely automatic, and thus suitable for telemetry. Of course, we could always consider using certain automatic methods (like that of Bacchus) which have been tried on the Earth and which have failed... precisely because of the atmospheric turbulence.

6. Relative Equatorial Astrometry

In terrestrial observatories, equatorial astrometry occupies a place of particular importance, despite the fact that it is currently out of favor for non-scientific reasons.

This is equally true for the observation of double stars (which, as we have just seen, give access to stellar masses), parallax measurements (which furnish the distances, and therefore the absolute luminosities, of the stars), and the determination of proper motions (which are connected with the motion of our Galaxy, and its dynamical and physical evolution).

Parallaxes and proper motions depend on the comparison of photographic plates, taken under identical conditions at intervals of months or years. The limitations in this field are chiefly due to the lack of old plates; but the limiting magnitude also plays a role. Is it entirely out of the question that a lunar observatory, working for many years, might redouble the indispensable work performed by terrestrial astrometry: taking the measure of our Galaxy?

7. Absolute Astrometry

We have already pointed out the problem posed by refraction in the establishment of star catalogues – namely, that refraction makes it very difficult to match up the northern and southern hemispheres. This matching could be done without great difficulty (with a judicious selection of comparison stars) by means of a satellite.

Thus we have seen in this chapter, as in the preceding one, that photometry and astrometry can benefit from space research, even in the ordinary wavelength region of classical astronomy.

However, we must realize that there are very difficult technical problems awaiting the investigators who will go into this line of research. They will have to point their instruments with a precision at least equal to the dimensions of the diffraction image, throughout the observations; and they will have to avoid confusion between the faint objects being observed and other, neighboring objects. For the fainter the magnitude, the smaller the mean separation between two objects on the sky. One cannot, therefore, gain indefinitely in magnitude, without making greatly increased demands on the pointing accuracy! Besides, either some difficult observations will have to be automated, or some astronomers will have to be put into orbit – it is hard to say which would be more of a problem. And finally, the apparent motion of a heavenly body with respect to the observer is simple at terrestrial observatories: it is a rotation about the Earth's axis. In a space station, or on the Moon, the situation is more complicated!

It is obvious that, at the moment, we are dealing only in speculation. But it is equally obvious to me that, thirty years from now, space astrometry will be a reality. Perhaps even sooner? The rapidity of progress in this field has always been surprising, ever since, more than ten years ago, the first Sputnik inaugurated the space age.

CHAPTER III

BEYOND ULTRAVIOLET OPACITY

Opening up the spectrum to astronomers by observations from space vehicles means, first and foremost (this will be a long chapter!), the use of measurements made at short wavelengths. But why go chasing the ultraviolet? There are several reasons for doing so.

1. The Ultraviolet is a Good Thermometer

First of all, the ultraviolet is a good thermometer. We have already mentioned that the ultraviolet (and, *a fortiori*, X- and gamma rays) give the best possible information concerning the temperature, to the extent that the measured radiation is of thermal origin. But this remains to be proved!

We know that the energy per unit frequency between v and $v+dv$, radiated by a heated body (black body) of temperature T, is

$$F_v = \frac{2\pi h v^3}{c^2} \frac{1}{e^{hv/kT} - 1},$$ (9)

in ergs per second per square centimeter of radiating surface. If we are interested in the energy per unit wavelength (instead of unit frequency) we have

$$F_\lambda = \frac{2\pi h c^2}{\lambda^5} \frac{1}{e^{hc/k\lambda T} - 1},$$ (10)

making use of the well-known formula $v = c/\lambda$.

In these two equations, the constants have the values indicated in Table I of the Appendix. The wavelengths λ are expressed in centimeters, and the temperatures T in degrees Kelvin (absolute temperatures).

The asymptotic behavior of the energy flux is simple: if λ is large, F is proportional to T (the Rayleigh-Jeans law); but if λ is small, F is proportional to e^{-KT} (Wien's law), where K increases with increasing $v = c/\lambda$ and therefore with decreasing λ. Thus it is $\log F$ that is proportional to T. It is therefore at short wavelengths that the observed energy is the most sensitive measure of the temperature.

Let us elaborate this point. First of all, we can see in Figures 31 and 32, which represent F_v and F_λ respectively, that the asymptotic laws are valid over rather broad regions. Roughly speaking, these regions are separated by a rather narrow frequency (or wavelength) zone centered on the maximum of the curve in question. This maxi-

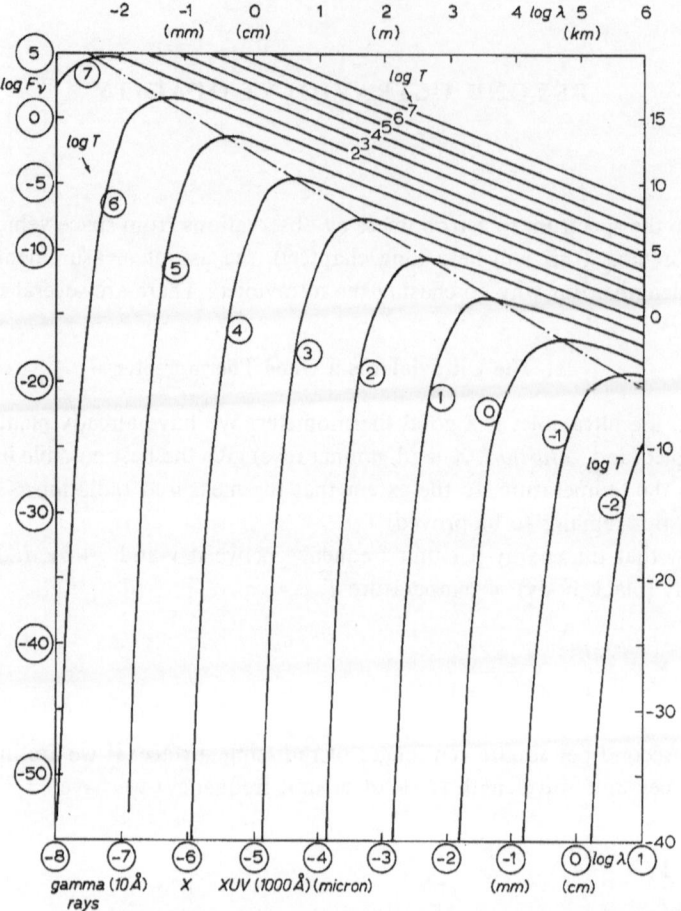

Fig. 31.   Planck function F_ν. – The circled values on the curves refer to the ordinate and abscissa
scales whose values are also circled. The same quantity appears on the left and right *ordinate*, and
similarly on the upper and lower *abscissa*.

mum (Wien's law) occurs at

$$\lambda_{\text{max}, \lambda} = 0.289\,75/T \quad \text{for } F_\lambda, \tag{11}$$

and at

$$\lambda_{\text{max}, \nu} = 0.509\,90/T \quad \text{for } F_\nu. \tag{12}$$

At a given temperature T, the spectrum is therefore a 'poor' thermometer if $\lambda > 10\lambda_{\text{max}}$,
and a 'good' thermometer if $\lambda < \frac{1}{10}\lambda_{\text{max}}$ (for $T = 5000°$, $\lambda_{\text{max}} \sim 0.6\,\mu$; but for $T = 10^6$, we
have $\lambda_{\text{max}} = 30$ Å).

Moreover, at a given wavelength the spectrum is more sensitive for low temperatures than for those above the Wien temperature given in Table VII.

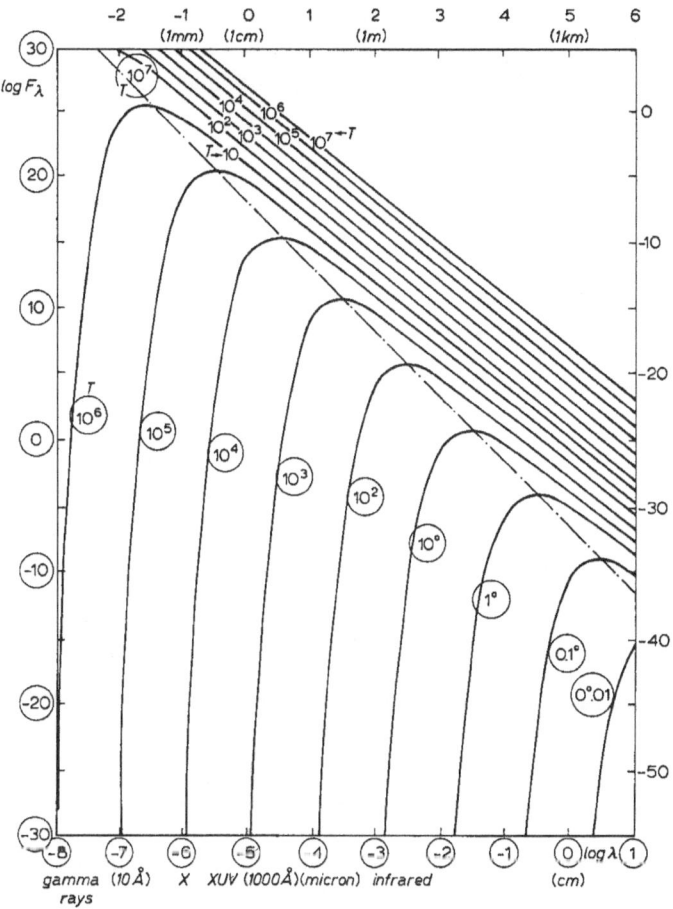

Fig. 32. Planck function F_λ. – The circled values on the curves refer to the ordinate and abscissa scales whose values are also circled. The same quantity appears on the left and right *ordinate*, and similarly on the upper and lower *abscissa*.

TABLE VII

Temperature of the Wien maximum

$\lambda =$			$T = 5 \times 10^7$	$T = 3 \times 10^7$
1 Å	$=$	10^{-8} cm	5×10^7	3×10^7
10 Å		10^{-7} cm	5×10^6	3×10^6
100 Å		10^{-6} cm	5×10^5	3×10^5
1000 Å		10^{-5} cm	50000	30000
5000 Å		5×10^{-5} cm	10000	6000
1 μ		10^{-4} cm	5000	3000
10 μ		10^{-3} cm	500	300
100 μ		10^{-2} cm	50	30
1 mm		10^{-1} cm	5	3

In the radio region, the Rayleigh-Jeans law is valid at all temperatures presently of astrophysical interest.

But the spectrum will be better and better as a thermometer, the smaller λ becomes: and the sensitivity of this thermometer can be precisely defined.

The sensitivity of the energy flux to a temperature fluctuation is

$$y_v = dF_v/dT \quad \text{or} \quad y_\lambda = dF_\lambda/dT , \tag{13}$$

depending on which variable one wants to use – the choice is determined by the experimental procedure, and by the properties of the detecting and dispersing systems.

These two quantities can easily be calculated. We find

$$\Psi = \frac{1}{F_\lambda} \frac{dF_\lambda}{dT} = \frac{1}{F_v} \frac{dF_v}{dT} = e^{hv/kT} \frac{hv}{kT^2} \frac{1}{e^{hv/kT} - 1} , \tag{14}$$

a quantity that obviously depends upon v (or λ) and T.

Figure 33 is a graph of this variation, taking as variables the logarithm of the wavelength and the temperature (which has more immediate significance than the quantity $x = hv/kT$). Note, however, that the quantity $T\Psi$ is easier to use, because it depends only on x:

$$T\Psi = x/(e^{-x} - 1). \tag{15}$$

We note that Ψ, defined as the 'sensitivity', corresponds to a logarithmic error and is expressed in 'exps' per degree*, similar to 'magnitudes per degree'.

This definition is *independent* of the measurement procedure. If we had expressed the sensitivity in ergs per degree, we would have had to use the quantities $y_v = \Psi F_v$ or $y_\lambda = \Psi F_\lambda$ per unit frequency or per unit wavelength.

The graphs of Figure 31 (F_v) and Figure 32 (F_λ) aid us in transforming Ψ to y_v or y_λ.

From the experimenter's point of view, what counts is the amount of information contained in the measurement. If, for example, we are using a photographic plate, whose response is assumed to be reduced to the number of photons N_v received per square centimeter of the plate, what happens? If the dispersing system is a grating, the dispersion is proportional to the wavelength; the information is therefore proportional to the number of photons of wavelength between λ and $\lambda + d\lambda$ (per unit wavelength interval). The number of photons is related to the energy by the formula:

$$N_v = F_v/hv \quad \text{or} \quad N_\lambda = F_\lambda/hv . \tag{16}$$

In the present case, the sensitivity should therefore be defined by a quantity

$$z = \frac{dN_v}{dT} = \frac{1}{hv} y_v . \tag{17}$$

---

* An 'exp' is the difference between two measurements whose ratio is equal to e; a 'dex' is the difference between two measurements whose ratio is equal to 10. We have: 1.0857 magnitude = 1.0000 exp = 0.43429 dex; 2.5000 magnitudes = 2.3026 exp = 1.0000 dex; 1 decibel is one-tenth of a dex (used to express noise volume); 1 octave is equal to 0.30103 dex (used to measure frequency intervals).

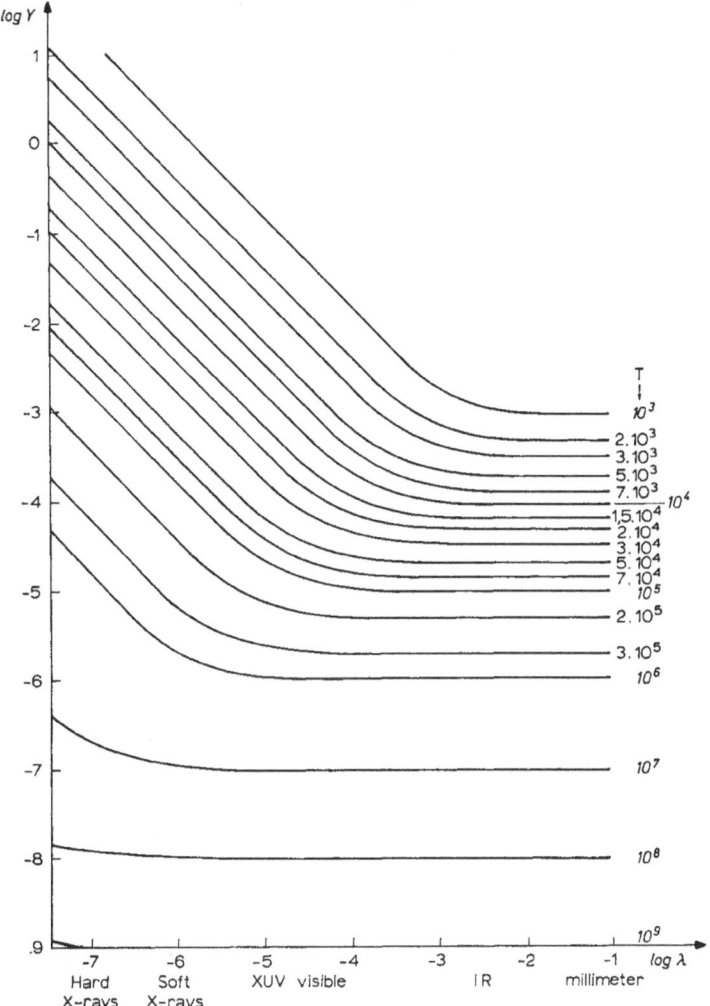

Fig. 33. The function $\Psi = (1/F)\,(\mathrm{d}F/\mathrm{d}T)$. This function measures the thermometric sensitivity of a given type of radiation.

In our opinion, a discussion of this nature – sometimes raised by the difficulties inherent in the choice of the energy variable (frequency or wavelength) – is quite academic, for the use of the quantity Ψ is perfectly clear and unique, whatever the spectral properties of the receiver.

Let us then return to our statement: the short wavelengths are an excellent cosmic thermometer. Indeed, Figure 33 shows us that a measurement made 'to a hundredth of an exp' (about a hundredth of a magnitude) corresponds to the values indicated in Table VIII for $T = 5000°$ and $T = 10^6$.

TABLE VIII

	$T = 5000$ K	$T = 10^6$ K	λ
$\Delta T = 1/100\ \Psi =$	0.056°	7.1×10^{2}°	10 Å (hard X-rays)
	0.56°	5.4×10^{8}°	100 Å (soft X-rays)
	1.8°	9.6×10^{8}°	1000 Å (X, UV)
	8.9°	10^{4}°	5000 Å (visible)
	4.4°	10^{4}°	10 μ (IR)
	50°	10^{4}°	10 cm
	50°	10^{4}°	10 m

In other words, for T around 10^6 degrees the hard X-rays have a thermometric sensitivity 15 times greater than visible light; and for $T \sim 5000$ K, the factor is on the order of 150. This fact is, as we have said, also very clear in Figures 31 and 32, which show that the isothermal black-body spectra are very close together at large wavelengths, but quite widely spaced at short wavelengths...

2. The XUV is a Good Detector of Solar Activity

So the XUV spectrum is certainly a good thermometer! Consequently, it is a good detector of solar activity. Let us consider an object (the Sun, a star) where hot, small, localized sources superimpose their effects upon the quiet background of the object as a whole. This is the case, for example, with the solar disk (quiet), upon which certain points are the site of rapid and violent flares.

Let us suppose, for example, that a fraction $s = 10^{-5}$ of the apparent surface area of the object in question (at the 'quiet' temperature T_q) is occupied by an 'active' zone of optical thickness distinctly greater than unity* and of temperature $T_a > T_q$.

The energy flux measured will be

$$F = sF_a + (1 - s) F_q \qquad (18)$$

per square centimeter of average radiating surface. Let us assume, for instance, that we are measuring only the total flux: for a star, this is always the case; for the Sun, it was the case at the time of the first rocket experiments, when there were no precise pointing systems.

If we put

$$F_a = \alpha F_q, \qquad (19)$$

we can write:

$$F = F_q[1 + (\alpha - 1) s] = F_q(\beta + 1). \qquad (20)$$

Clearly, α is a function of the wavelength. We have:

$$\alpha = \frac{F_a}{F_q} = \frac{e^{h\nu/kT_q} - 1}{e^{h\nu/kT_a} - 1}, \quad \text{and} \quad \alpha - 1 = \frac{e^{h\nu/kT_q} - e^{h\nu/kT_a}}{e^{h\nu/kT_a} - 1}. \qquad (21)$$

* We shall explain in the following section the meaning of this temporarily mysterious restriction.

Here too, we could obtain a general idea of the asymptotic behavior corresponding to the Rayleigh-Jeans and Wien laws. But the result is obvious:

If $\beta = (\alpha - 1)s$ is significantly less than unity, the active region will be difficult, if not impossible, to detect.

If β is on the order of unity, detection will be easy.

If β is much greater than unity, the measurement will represent *only* the active region.

Let us discuss some typical examples: first, activity of the chromospheric type $(T_a \sim 2T_q = 2 \times 10^{4\circ})$; next, activity of the coronal type $(T_a \sim 2T_q = 2 \times 10^{6\circ})$; and finally, the solar granulation $(T_a \sim 1.1\, T_q \sim 5500°)$.

Table IX gives the quantity $\alpha - 1$ as a function of wavelength.

TABLE IX

λ	$\alpha - 1$ 'Granulation'	'Chromosphere'	'Corona'
0.1 Å	1.17×10^{11361}	9.24×10^{31242}	2.69×10^{312}
1 Å	1.28×10^{1136}	1.98×10^{3124}	1.75×10^{31}
10 Å	3.99×10^{113}	2.69×10^{312}	1332
100 Å	2.30×10^{11}	1.75×10^{31}	2.053
1000 Å	12.68	1332	1.075
1 μ	0.3226	2.05	1.007
1 mm	0.1001	1.00	1.000

We can draw several conclusions from this calculation.

If the active region is small $(s = 10^{-n})$, it will be more difficult to detect in the corona than in the chromosphere. Thus if $n = 10$, the active region will begin to influence the overall observations at hard X-ray wavelengths for the corona, and at XUV wavelengths for the chromosphere. In any case, the detectability increases considerably as the wavelength decreases. For the granulation $(s = \frac{1}{10}$ is typical), it is also in the XUV region that detection becomes easy; but then the opacity of the chromosphere is too great for the radiation to be influenced by the (photospheric) granulation. At least we are sure that in the XUV region, only the hot portions of the chromosphere (spicules or interspicules? $s = 10^{-4}$?) will affect the observations.

Figure 34, which represents an overall view of the solar spectrum, is a good illustration of this fact.

3. The Opacity of Astrophysical Media is High at Short Wavelengths

This is another essential reason for using the ultraviolet to study the outer regions of stars or of the Sun.

It is, in fact, quite remarkable to realize that the most penetrating wavelengths are those of the visible spectrum. They penetrate the eye, but they are not alone in doing

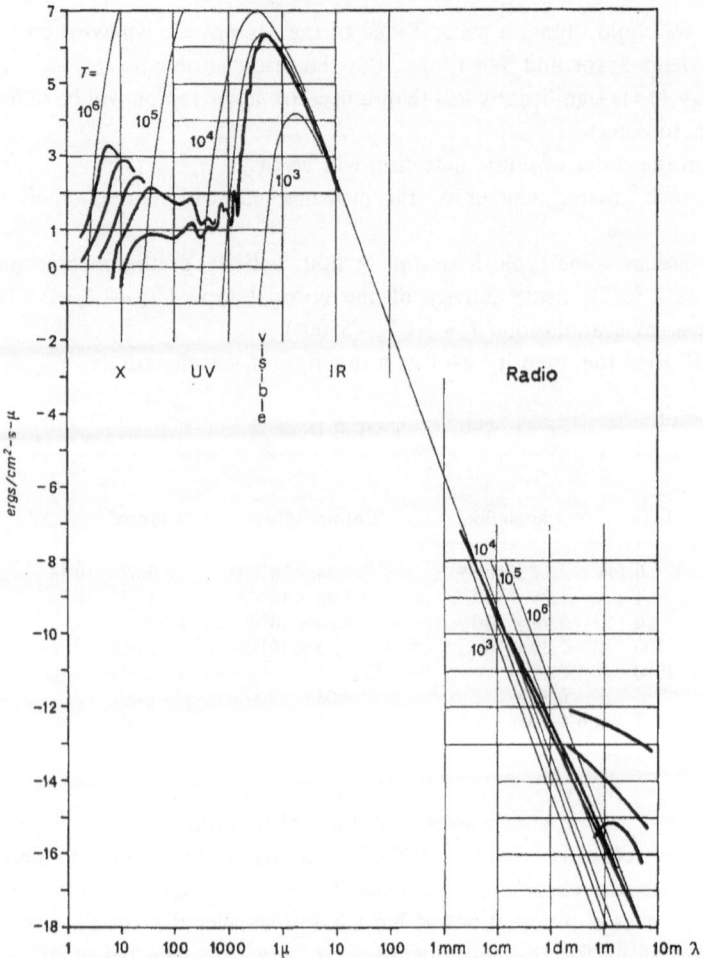

Fig. 34.   The solar spectrum (after a Ball Brothers leaflet). – *Ordinate:* logarithm of the radiative energy flux. The light curves represent Planckian distributions; the heavy curves represent the observed distribution. In the radio and XUV regions, note that the spectrum of the active chromospheric and coronal regions dominates the spectrum of the 'quiet' Sun.

so: the ultraviolet and X-rays do the same thing. They are, however, the only wavelengths that produce a reaction in the retinal cells – hence their name of 'visible' spectrum. But we have also seen that they penetrate the terrestrial atmosphere better than other wavelengths. This is no accident: air molecules resemble those of the eye to a certain extent, in their relative simplicity. It is also no accident that these are the most penetrating wavelengths in stellar atmospheres. The properties of the absorbing atoms to be found there are similar to those of the simple molecules of our atmosphere and our eyes.

It is not our intention to discuss in any detail the problem of the opacity of stellar atmospheres. Spectral studies have shown that the principal constituent of the universe is hydrogen, in all its forms: hydrogen molecules (H_2) in the interstellar medium and the atmospheres of cold stars, and even hydrides and hydrates (CH, OH, OH_2, NH_3, CH_4) in the planets; negative hydrogen ions in stellar atmospheres of low and medium temperature; atoms of neutral hydrogen in stellar atmospheres of low and medium temperature, and in the HI regions of the interstellar medium; ionized hydrogen (i.e., protons) in hot stars, in the central regions of all stars, and in the HII regions of the interstellar medium... However, atoms much less abundant than hydrogen, but individually very opaque, can make large contributions to the opacity in certain regions of the spectrum. In particular, this is the case for many metals and ionized metals and for various molecules, principally oxides (like carbon monoxide CO, titanium oxide TiO, strontium oxide SrO, etc.). Moreover, scattering by free electrons is an obstacle to the penetration of the hottest media, while scattering by solid grains is an obstacle to the penetration of the coldest media.

TABLE X

Chemical composition of the Sun

(logarithms of the relative abundances)

H	10.50	Ca	4.54	Y	1.70
He	9.71	Sc	1.30	Zr	1.15
Li	0.04	Ti	3.08	Nb	0.80
Be	0.84	V	2.62	Mo	0.80
C	7.22	Cr	3.40	Ru	0.32
N	6.48	Mn	3.30	Rh	−0.13
O	7.46	Fe	5.14	Pd	−0.23
F	3.2	Co	3.20	Ag	0.46
Na	3.94	Ni	4.45	Cd	0.16
Mg	5.90	Cu	2.00	In	−0.22
Al	4.98	Zn	2.30	Sn	0.55
Si	6.00	Ga	1.25	Sb	−1.08
P	3.84	Ge	0.99	Ba	1.00
S	5.80	Rb	0.98	Yb	0.78
Cl	4.75	Sr	1.20	Pb	0.14
K	3.16				

In the particular case of a star like the Sun, we can determine, from tables of opacity and of chemical composition (Table X), the layer of the star in which the observed radiation originates.

If we calculate the radiative intensity I_v as the integral over depth in the atmosphere of the contribution of the different emitting layers, we can write

$$dI_v = \varepsilon_v(x)\, e^{-\tau_v(x)}\, dx, \tag{22}$$

where $\varepsilon_v(x)$ is the specific radiative emissivity in the layer of geometrical depth x;

τ_v is the optical thickness

$$\tau_v = \int_x^\infty \kappa_v \rho \, dx \tag{23}$$

of the matter traversed between the point of emission and the observer; and ρ is the density.

If we assume Kirchhoff's law, the ratio ε_v/κ_v is equal to the function $B_v(T)$, which represents the specific radiative intensity of a black body of temperature T. This condition is sufficiently well satisfied in the continuous spectrum; in any case, the specialists in the theory of atmospheres know how to calculate the ratio ε_v/κ_v in a more precise fashion.

Thus we have

$$I_v = \int dI_v = \int_{-\infty}^{+\infty} B_v(x) e^{-\tau_v} \kappa_v \, dx; \tag{24}$$

or, if we take τ_v as the variable of integration,

$$I_v = \int_0^\infty B_v(\tau) e^{-\tau_v} \, d\tau_v, \tag{25}$$

an equation familiar to astrophysicists.

If B_v is a linear function of τ_v (we can show that this is so, in the first approximation), we can write

$$I_v = \int_0^\infty (a + b\tau_v) e^{-\tau_v} \, d\tau_v = a + b = B_v(\tau_v = 1). \tag{26}$$

This means (a fact which has been known since Eddington's time) that it is just as if the radiation originated in the layer of optical depth $\tau_v = 1$. The regions whose optical thickness is significantly less than unity are transparent, and do not affect the observed radiation; the regions whose optical thickness is significantly greater than unity are opaque, and hide the deeper layers from the observer.

It is customary to take the optical depth τ_0 at 5000 Å, in the spectral region of greatest transparency, as the unique parameter of depth.

The layers whose optical depth at wavelength λ is on the order of unity, have an optical depth τ_0 equal to:

$$\tau_0 = \int_0^1 \frac{\kappa_0}{\kappa_v} \, d\tau_v. \tag{27}$$

For a given star – the Sun, for example – we know the variation of the physical quantities (temperature, pressure) as a function of τ_0, and we know the relationship $x(\tau_0)$ between the geometric height and the optical depth: this is what we call the 'model' of the atmosphere. Figure 35 gives a schematic outline of the currently-accepted model for the solar atmosphere, and shows the various layers (whose separation is not, however, always very clear): from the interior to the exterior, they are the photosphere, the chromosphere, and the corona.

At a given wavelength, the radiation originates in the layer $\tau_v = 1$. Using Equation (27) and our knowledge of the opacity, we can derive τ_0. In the case of the Sun, Figure 36 presents the results of such a calculation, carried out by the author. Note

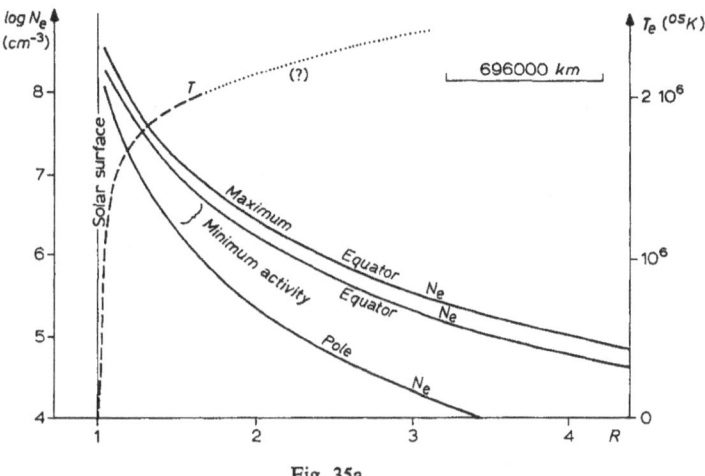

Fig. 35a.

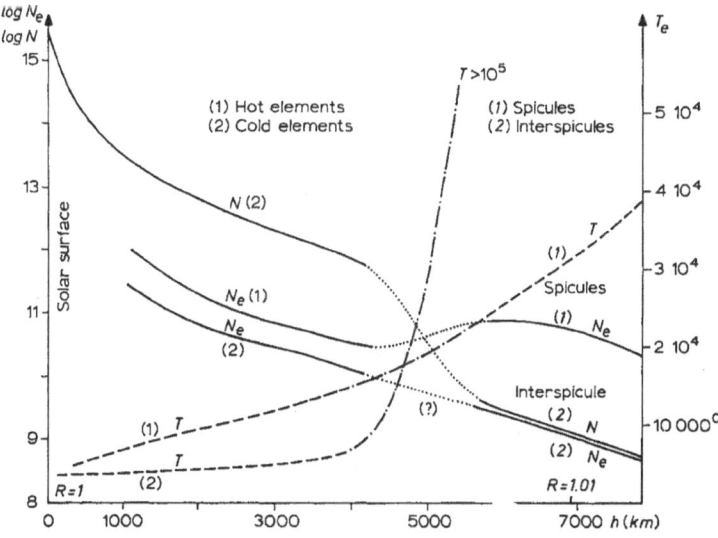

Fig. 35b.

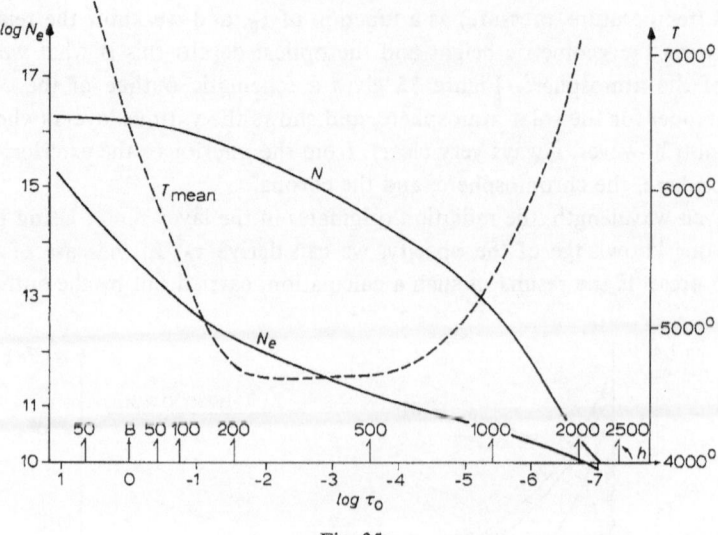

Fig. 35c.

Fig. 35. Model of the outer layers of the Sun. – *Dashed line:* electron temperature; *solid line:* density and electron density. (a) corona; (b) chromosphere; (c) photosphere. *Abscissa:* (a) height above the center of the Sun – unit: 1 solar radius; (b) height above the 'surface' – unit: 1 km; (c) logarithm of the optical depth at $\lambda = 5000$ Å.

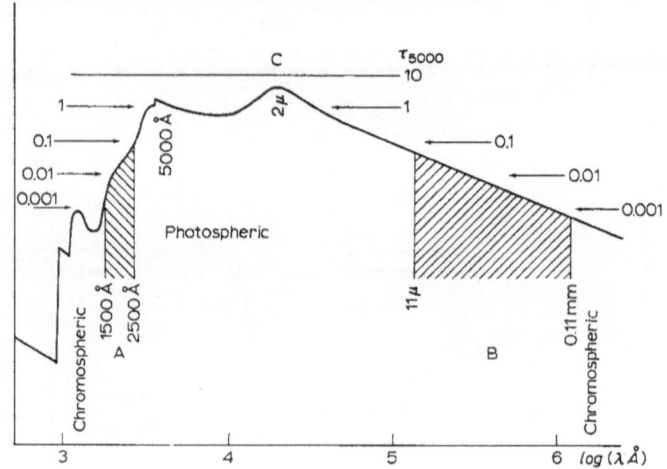

Fig. 36. Origin of the different parts of the continuous spectrum of the Sun.

that De Jager (1963) has independently located (at 15 μ) the infrared region, almost devoid of lines, which corresponds to the 1800 Å ultraviolet region.

By comparing this diagram with the solar model furnished by Figure 35 (or, to be exact, with the detailed tables on which this figure was based), we see which regions of the spectrum are formed in the chromosphere, and which in the corona. Then it becomes apparent that the near-ultraviolet radiation originates in the photosphere, while the XUV radiation originates in the chromosphere and the X-radiation in the corona. The latter wavelengths are therefore ideal media for exploring the outer layers of the Sun, which are opaque to them while remaining transparent to the visible spectrum, except in certain spectral lines which are very difficult to study.

It is true, as we see, that the infrared and radio radiation also come from the outer layers. But these wavelengths are poor thermometers, not very sensitive to thermal solar activity (we shall see that for non-thermal radiation, the shoe is on the other foot!). This is why *the study of short-wavelength radiation is essential for a knowledge of the outer layers of the Sun, especially the active parts – that is, the chromosphere and the corona.*

One further comment is necessary at this point. The opacity of certain active regions can be small at a given wavelength; this is the case, for example, with chromospheric flares that are not observed in the visible continuous spectrum. In the ultraviolet, these regions become optically thick: they are thus observable. This effect is added to the 'sensitive thermometer' effect studied above, and reinforces the opinion we expressed in the preceding section: the XUV spectrum is a good – a very good – detector of solar or stellar activity.

4. The Most Interesting Spectral Lines are Found in the X-Ray and XUV Spectra

There is almost a superfluity of good reasons for a careful study of short-wavelength radiation. Among these, it would be unfair not to mention the importance of the spectral lines that can be observed in the ultraviolet.

We know that the spectrum of a star (a hot body) consists mainly of a continuous spectrum comparable to that of a black body at temperature T, represented in Figures 31 and 32. But the atoms in the atmosphere of this star are 'quantized' systems, which means that their internal energy cannot take on any arbitrary value. We know that an atom is made up of a nucleus (positive), about which there gravitates a train of electrons (negative).

Each electron can occupy an orbit, an 'energy level'. The internal energy of the atom can change only if an electron jumps from one orbit to another; thus to each possible electron configuration there corresponds an energy. In such a jump or transition between two energy levels, a loss of energy can take place. This energy is then released in the form of a photon of energy

$$E = E_{\text{high}} - E_{\text{low}}, \tag{28}$$

where E_{high} and E_{low} are the energies of the upper and lower levels, respectively, of the transition in question. This photon has a wavelength λ and a frequency v, and we have

$$E = hv = hc/\lambda, \tag{29}$$

where h is Planck's constant. All this has been well known for fifty years...

Inversely, we know, a photon of energy $h\nu$ can be absorbed by the atom in question. The absorption of the photon's energy is translated into the transition of an electron to a higher energy level.

Every atom and ion has well-defined energy levels. To each transition between two energy levels there corresponds a line, quite localized in the spectrum – sometimes an emission line, sometimes an absorption line. What is the purpose of precise measurements of the radiation in spectral lines? Naturally, to learn the physical conditions in the solar or stellar atmospheres or, more generally, in the media responsible for the observed lines. Also, to measure the abundance of the corresponding elements – iron, calcium, etc. – with respect to hydrogen, the reference element.

Now a given line corresponds to the transition between a level n and a level m. It is stronger if the level n (the initial level of the transition) has a higher population – that is, if a larger number of atoms of the species in question (neutral iron, for example) are found in the given energy level. In fact, it is therefore the number of atoms in the level n that can be derived from the measurements (sometimes with great difficulty) and not the *total* number of atoms (of iron) under consideration. To obtain the latter, which is the only really interesting piece of information concerning the abundance of the atom in the astrophysical medium under consideration, we must know the distribution of the atoms over all energy levels – and particularly, of course, over the most highly populated levels. Now, the population N_n of a level n of energy E is equal to

$$N_n = (g_n/U)\, N e^{-E_n/kT} \tag{30}$$

(the Boltzmann equation), where E_n is the energy difference between the level n and the ground level of minimum energy (an atom cannot go below the ground level – the level that all atoms would occupy at a temperature of absolute zero); g_n is the statistical weight of the level n; U is the 'partition function' of the atom; and N is the total number of atoms in the various levels, including the ground level and all the excited levels. (Note that the number g_n/U is of order of magnitude unity, or nearly so.)

We see that the ground level $(E_n = 0)$ has a far greater population than a level n such that the quantity $e^{-E_n/kT}$ is significantly less than one (or, equivalently, such that $E_n/kT > 1$).

To reach levels that make a significant contribution to the abundance, it is therefore really necessary to reach the ground level and to measure the lines – called 'resonance lines' – which correspond to transitions arising from this level. Now, the ground level and the nearest excited levels are often widely separated. *Therefore resonance lines often have rather short wavelengths.* Hence, once more, the interest of this wavelength region, Table XI gives a list of the most interesting transitions in the X-ray spectrum. It is complemented by the list of XUV emission lines in Table XIII (Chapter IV, Section 2). Finally, we must not forget the more 'classical' lines, like the MgII doublet at 2796 and 2803 Å.

TABLE XI

Lyman continuum and important lines in the extreme ultraviolet and X-ray regions

λ(Å)	Number of photons emitted per second, per cm² of the solar surface	Identification
912–900	30 × 10⁸	Hɪ Ly Cont
900–840	32 × 10⁸	Hɪ Ly Cont
885–865	18 × 10⁸	Hɪ Ly Cont
835–833	5.7 × 10⁸	Oɪɪ, Oɪɪɪ
790–788	3.0 × 10⁸	Oɪᴠ
780–770	3.7 × 10⁸	Neᴠɪɪɪ
704–702	2.8 × 10⁸	Oɪɪɪ
630–625	9 × 10⁸	Oᴠ, Mgx
610	4.4 × 10⁸	Mgx
584	10 × 10⁸	Heɪ
554	3.2 × 10⁸	Oɪᴠ
553–501	1 × 10⁸	Heɪ, Oɪɪɪ
500	2.9 × 10⁸	Sixɪɪ
465	1.9 × 10⁸	Neᴠɪɪ
368	4.1 × 10⁸	Mgɪx
360	3 × 10⁸	Fexᴠɪ
335	3.1 × 10⁸	Fexᴠɪ
323–310	2 × 10⁸	
304	26 × 10⁸	Heɪɪ Lyα
285–282	4.5 × 10⁸	Fexᴠ
280–260	5 × 10⁸	
259–228	15 × 10⁸	Heɪɪ
228–200	20 × 10⁸	Ly Cont
200–166	20 × 10⁸	

(See also Table XIII, Chapter IV, Section 2.)

5. And Gamma Rays?

At extremely short wavelengths, gamma radiation is the extension of X-radiation. In general, it consists of photons whose energy is greater than 10^5 eV, and whose wavelength is shorter than a tenth of an angstrom.

Unlike radiation of wavelength greater than about 1 Å, which we have been discussing until now, gamma radiation is extremely penetrating; and almost all astrophysical media are transparent to it.

The importance of the study of this radiation is closely connected with its origin. The emission of gamma rays is closely associated with nuclear reactions and, as their origin can be located in distant regions of time or space, it has an obvious cosmological significance.

It is therefore quite natural to see a field of 'gamma astronomy' developing. Ground-based studies can perhaps reach the most penetrating gamma rays; but these are always perturbed by the secondary gamma rays produced in the atmosphere by high-energy incident particles. In any case, one has to go to enormous energies if the atmosphere is not to be completely opaque to gamma rays, and this is why gamma astronomy is essentially a branch of space research. We shall return to this subject a little later, in Chapter IV, Section 2.

But in this chapter of prospectives, perhaps it is appropriate to summarize the essentials of what could be obtained with gamma astronomy.

The *formation processes* of gamma rays are of several kinds.

First of all – like X-ray, ultraviolet, or other emissions – gamma rays arise from transitions between energy levels. But it is no longer a question of atoms (whose energy levels are very close together) but of atomic nuclei or of fundamental particles (heavy particles: protons, neutrons, hyperons; or low-mass particles: electrons, positrons, neutrinos). When these particles collide, some of their kinetic energy is converted into the energy of gamma photons.

Gamma rays also arise – and this is probably far from negligible – from the annihilation of particles and the corresponding anti-particles (particularly positrons and electrons).

Now, the chances are good that these different processes for the creation of gamma rays can take place in astrophysical media.

Their effects are easily differentiated. Most of the electromagnetic interactions give rise to continuous spectra, and the energy distribution measured in the hypothetical spectrum of a source of gamma rays should make it possible to determine the relative importance of the various interactions at work. Often, too, we are dealing not with a continuous spectrum but with a gamma 'line', comparable to the spectral lines at ordinary wavelengths. Thus an important line at 0.51 MeV (see Table III, Part I, Chapter II, for the relation between wavelengths and energies expressed in MeV) corresponds to the annihilation of positrons arising from the evolution of Π^+ mesons. Another important transition at 2.23 MeV is due to the formation of deuterons, and requires a flux of neutrons. Other lines of energy less than 10 MeV are due to the de-excitation of various nuclei (particularly C^{14} and O^{16}, in the astrophysical case).

In the case of the *Sun*, we hope to measure intense fluxes of gamma rays from flares. Thus a measurement of the line at 2.23 MeV would give us an idea of the neutron flux in a flare. Measurements of the continuous flux, on the other hand, would be associated with meson decay, and would provide some information concerning the distribution of the energetic particles accelerated during a flare.

Mechanisms for the production of gamma rays are also present in the *interstellar* or even the *intergalactic* medium. At high energies, there may be decay of Π mesons produced by interaction between the high-energy particles and the gas. Measurement of the resulting gamma rays would indicate the nature of the cosmic rays present in the Galaxy. Note that analogous mechanisms can take place in between the galaxies of the Metagalaxy, where unfortunately the physical conditions are too poorly known to make a valid prediction possible.

The *discrete sources of gamma rays* are no doubt of greater interest. Since the production of gamma rays is more or less connected with the production of high-energy particles, the gamma-ray sources are also sources of high-energy particles, and gamma astronomy is a method for detecting them. They are sources of radio radiation (synchrotron radiation, caused by the braking of high-energy electrons at relativistic velocities in magnetic fields, produces radio radiation at very long wavelengths) as well as of gamma rays. But a gamma source is not *necessarily* a radio source. It is clear that quasars (and other objects called 'quasi-stellar', no doubt because they have nothing whatsoever to do with stars) must be intense sources of gamma rays.

It would be very useful to give in detail the mechanisms for production of gamma rays. But ordinary astronomy deals with low energies. Gamma rays belong to the field of high energies, and a complete course on interactions of all kinds between high-energy particles should preface any advanced discussion. The author (whose laziness and ignorance do not allow him to write such an introduction in a few pages) will be forgiven if he limits himself to the above considerations. We shall see in the next chapter, Section 2, that gamma astronomy has not yet made any great contribution to the progress of astronomy as a whole. But it would not be surprising if, a few years from now, these few paragraphs were to be expanded into an entire volume in this series.

CHAPTER IV

ASTRONOMY AT SHORT WAVELENGTHS

1. Short-Wavelength Radiation from the Sun

First among the discoveries of space astronomy are certainly those concerning the radiation of the Sun at short wavelengths, for the reasons discussed in the preceding chapter. It could undoubtedly be said that the first significant observation was that obtained by Rense in 1952 (see Figure 37). His image of the Sun in the Lyman α line was no doubt of mediocre quality, but it was nevertheless the point of departure for an impressive series of projects in the ultraviolet and X-ray regions. Most of this research was carried out by scientists in the United States, and it is obviously impossible to give a complete and detailed historical description here. We shall merely attempt to synthesize the results obtained.

A. EXPERIMENTAL RESULTS

First of all, these experiments contributed *spectra* of the Sun, at different points on the solar disk. Then, to complement the preceding information, *images* of the Sun were obtained at various wavelengths corresponding to more or less 'monochromatic' bandwidths.

Figure 38 shows the principal characteristics of the solar ultraviolet spectrum. Schematically speaking, we can say that for wavelengths greater than 1800 Å there is an absorption spectrum, comparable to that obtained at visible wavelengths: this is still the photospheric spectrum. At wavelengths between 1800 Å and the Lyman discontinuity of hydrogen at 912 Å, as well as in the continuum at shorter wavelengths, we are dealing with chromospheric radiation. The lines appear in emission and the Lyman discontinuity, like the other Lyman lines, is characteristic of the neutral atom of hydrogen. One has to go to much shorter wavelengths in order to find a spectrum formed entirely in the corona; we can say that at wavelengths shorter than about 500 Å, most of the information obtained is principally concerned with the coronal layers – with the exception of certain very strong lines like those of neutral and ionized helium, which are obviously associated with the layers of the chromosphere. A very large number of spectral lines have been identified and measured; the continuous spectrum itself has also been measured rather completely. In the XUV, there are 92 neutral or ionized elements known! Figure 39 shows the principal measured characteristics of the continuous spectrum.

It is quite remarkable to note that center-to-limb measurements display variations in the behavior of different spectral regions. Figure 40, due to Bonnet, shows in

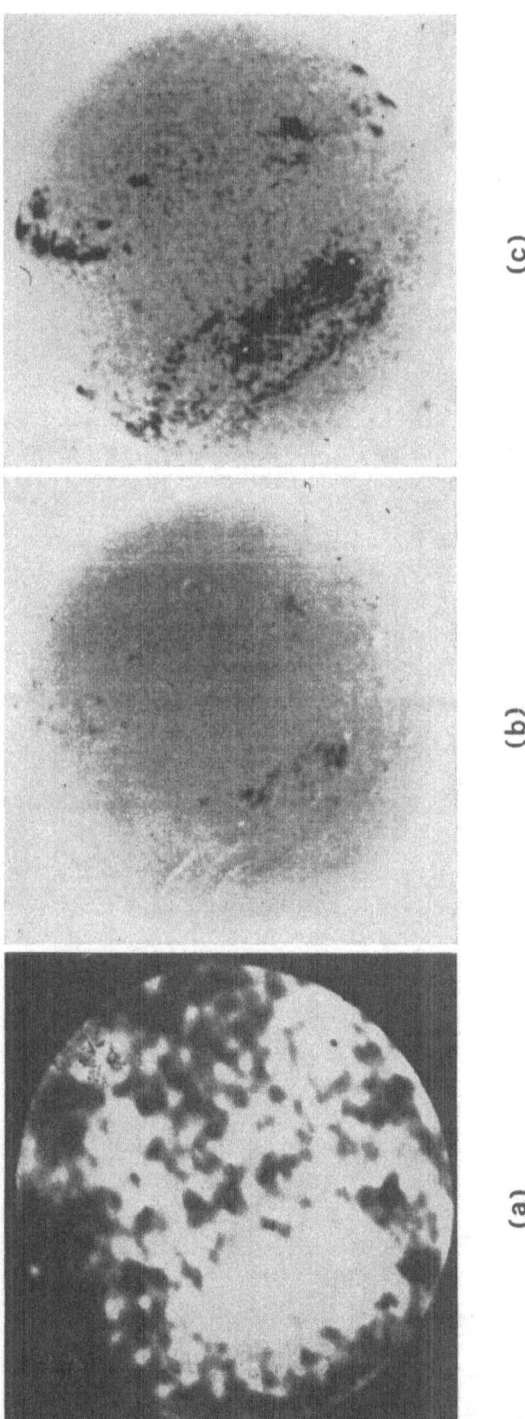

(a)

(b)

(c)

Fig. 37.   The first image of the Sun in Lyman α (a), obtained in 1952 by W. A. Rense; (b) image of the Sun taken on the same day in Hα; (c) image in the K line of ionized calcium, also taken on the same day. (Photos by W. A. Rense.)

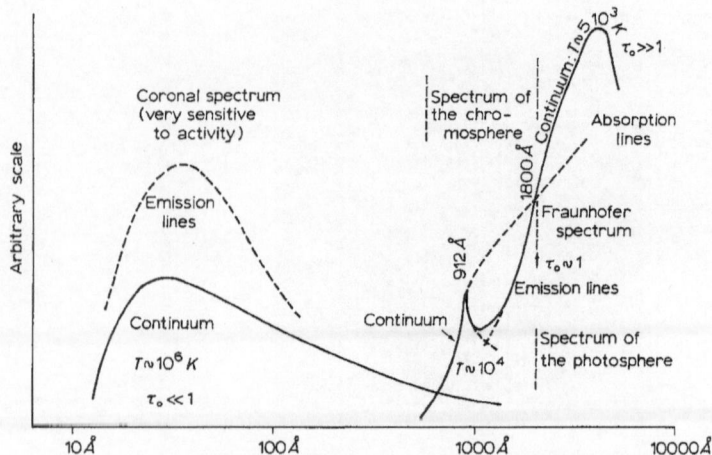

Fig. 38. Characteristics of the solar ultraviolet spectrum. Compare this figure with Figure 34, whose left-hand portion (XUV) is outlined here.

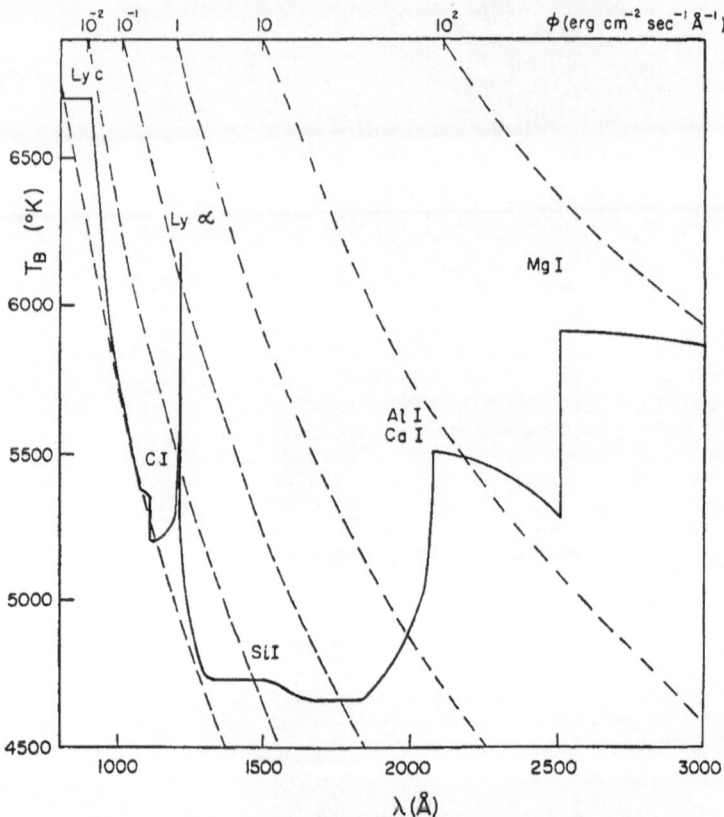

Fig. 39. Ultraviolet continuous spectrum (after Goldberg). – *Ordinate:* brightness temperature, or temperature of a black body that would radiate the same energy flux per unit surface area. Note the role played by the metallic continua, particularly that of Mg I and by the Lyman continuum (Ly C) of hydrogen.

particular that the center-to-limb variation is very different from one point to another of the continuous spectrum; thus the discontinuity (still not completely identified) at around 1800 Å practically disappears at the limb; we shall see below how results of this kind can be interpreted. In the X-ray region, a limb brightening reflects the temperature increase in the corona; since the spectrum is an "image" of the temperature distribution in the layers responsible for it, this circumstance corresponds to a temperature increasing outwards in the corona, revealed by an increase in opacity towards shorter wavelengths. We recall that Figures 36 and 38 summarize the origin of different portions of the solar spectrum.

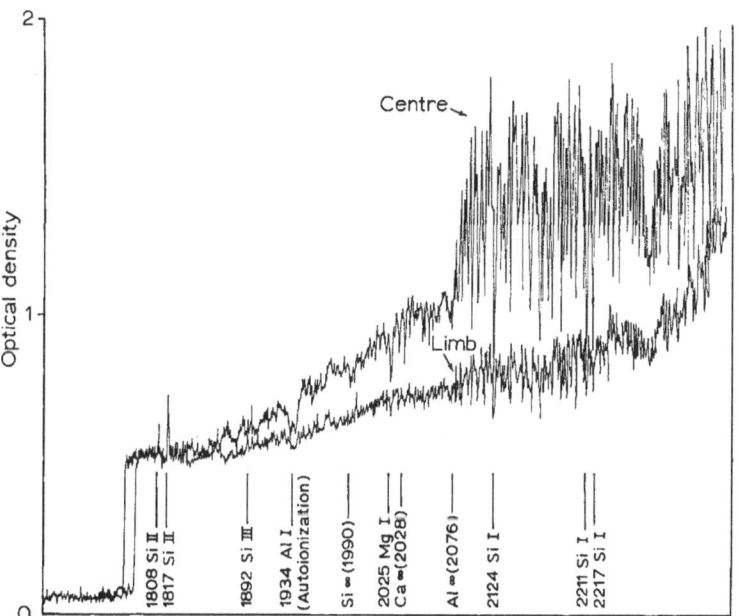

Fig. 40. Solar spectrum in the ultraviolet (after R. Bonnet). – *Left*, at the limb; *right*, at the center of the disk. Note the important discontinuity at around 2100 Å: theory shows that the observations cannot be explained by attributing this discontinuity to Al I.

Obtaining monochromatic photographs of the Sun is obviously an extremely important item; in these photographs, the active regions are more or less apparent depending upon the sensitivity of the chosen wavelength to the temperature or the density, whichever of these physical quantities is the determining factor in a particular active region. Many devices have made it possible to obtain such images of the Sun. Thus we have images of the Sun in the 304 Å line of He II, which strongly resemble the photographs obtained in the K line of calcium; images of the Sun in various regions of the X-ray spectrum (the 50 Å band, the 10 Å band, the 5 Å band: see Figure 41); images of the Sun in Lyman α and Lyman β; and images of the Sun in

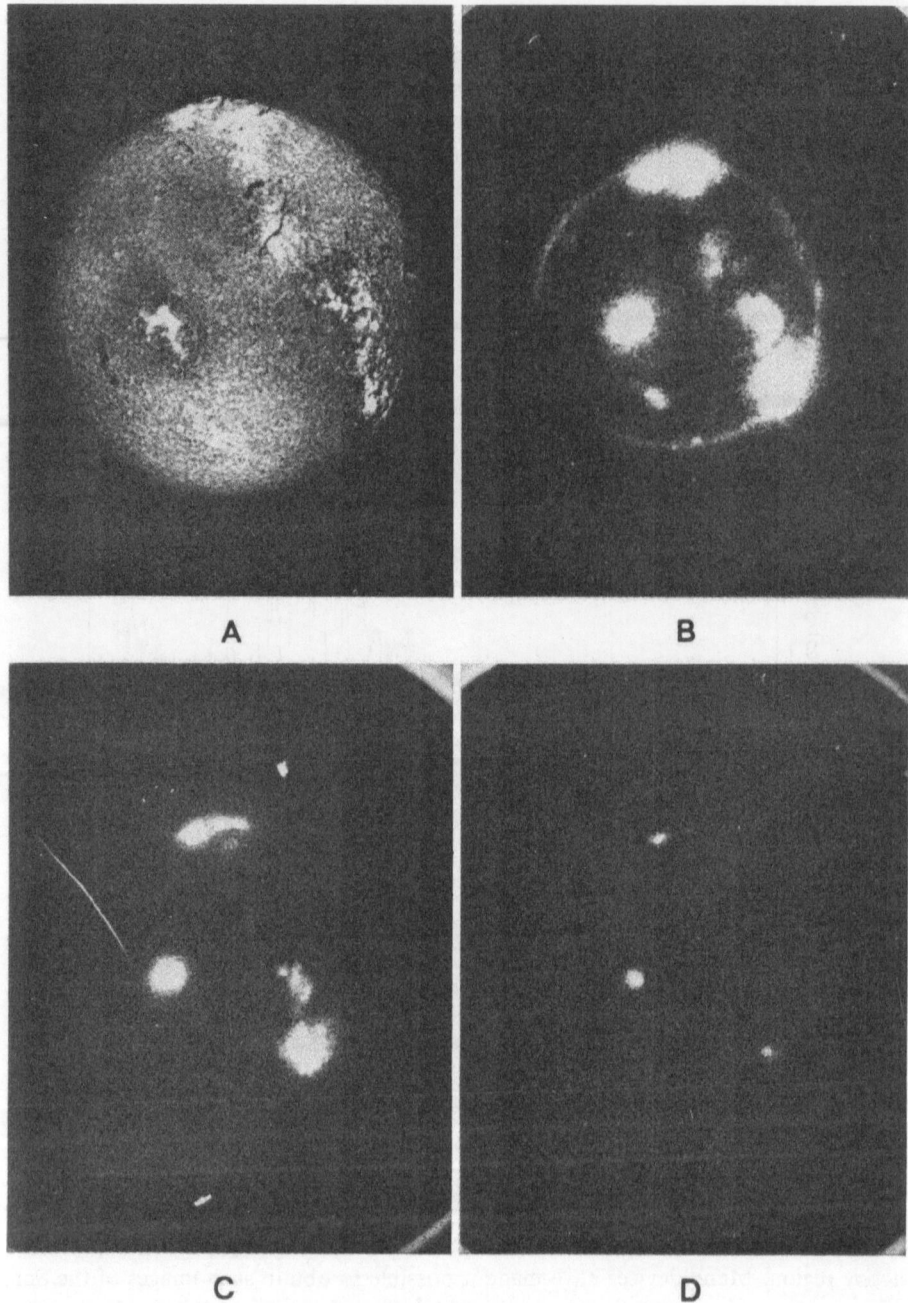

Fig. 41.   X-ray photographs of the Sun (NASA Goddard Space Flight Center). – A: Hα; B: λ44–60 Å
and 3–13 Å; C: λ8–20 Å; D: λ3–13 Å.

the continuous spectrum between 2000 and 3000 Å. The photograph in Figure 42, taken by Bonnet, shows the Sun in continuous radiation at a wavelength of 1980 Å. Note that the *plages* already appear at these wavelengths, even at the center of the disk, while in white light they appear only at the limb; this is due to the fact that by

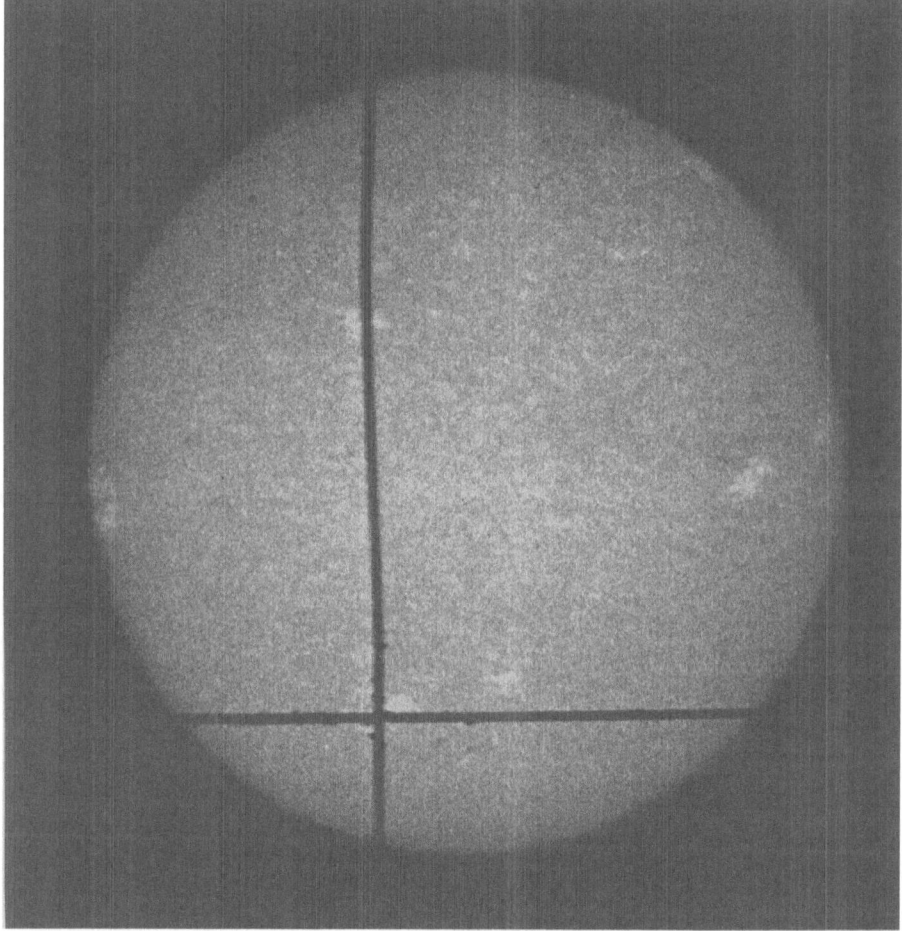

Fig. 42.   Photograph of the Sun in continuum light (λ: 1980 Å, photo by R. Bonnet). – Note the faculae (or *plages*) visible near the center: the chromosphere is already rather opaque at this wavelength.

changing wavelengths, we have gone higher up in the atmosphere. The interpretation, then, is simple: the plages are essentially important manifestations of chromospheric activity, but they are greatly weakened in the photosphere and even nonexistent at great depths.

The *variation in time* of the radiation has also been observed, of course. As we have said (Chapter III, Section 2), the ultraviolet radiation – and to an even greater extent, the X-radiation – is an excellent detector of activity; consequently, measurements made at short wavelengths should enable us to detect solar activity and measure it with precision (Figure 43). Figure 44 shows the variations obtained in the 2–8 Å

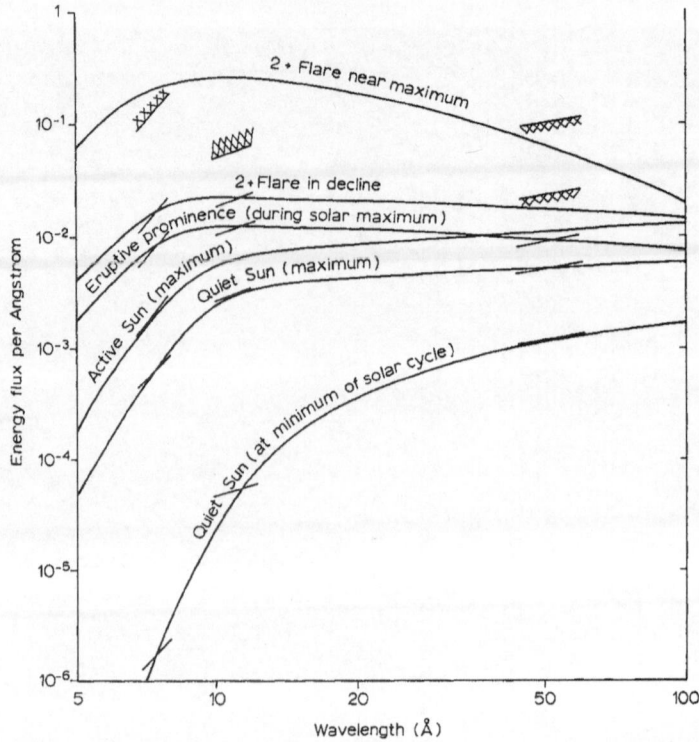

Fig. 43.   X-ray spectrum of the Sun and of various types of solar activity (after Goldberg).

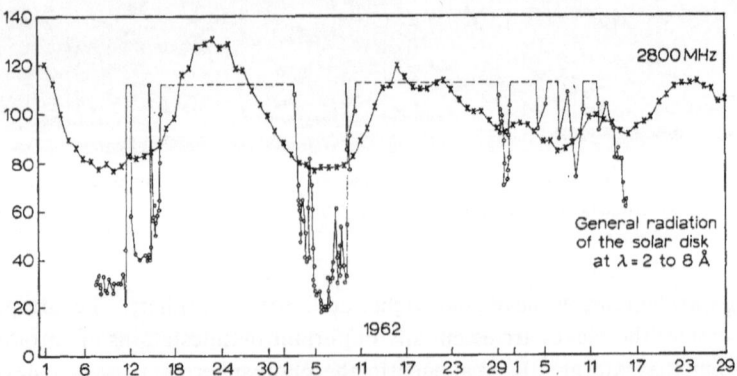

Fig. 44.   Fluctuations in the solar X-radiation (compared with radio radiation at 2800 MHz). (Measurements from Goddard Space Flight Center.)

region, compared with observations made at radio wavelengths, which also originate in the coronal regions. Note that the variations in the X-ray region are much more noticeable than those in the radio region, which is not at all surprising considering what we said in Chapter III, Sections 1 and 2, concerning the sensitivity of short-wavelength radiation to temperature fluctuations.

It is now clear that a very large number of problems concerning the active regions can be solved by analyzing the spectrum and images of the Sun at ultraviolet and X-ray wavelengths. We shall give below a few examples of the extraction of information from the observations (once again without pretending to give a complete exposé of the results).

B. INTERPRETATION

Concerning the *photospheric* layers, we can say that space research has not told us much that we did not already know. It is principally the continuous spectrum (in the visible region) that gives us information concerning the photospheric structures. On the other hand, the contributions of space research have been very important as regards the *transition zone between the photosphere and the chromosphere*. For in the near-ultraviolet region, we can say that the spectrum results from both the opacity and the model, and can therefore give information concerning both the opacity and the model. But since the theoretical data in the far infrared specify the model (because in the far infrared the absorbent – the negative hydrogen ion – is known), knowledge of the opacity is the most important information to be deduced from those portions of the spectrum that originate in the chromosphere-photosphere transition region. Thus, for example, the data obtained by Bonnet enable us – by a judicious interpretation of the discontinuity at 1800 Å – to derive certain characteristics of the chemical composition of the solar atmosphere; we know that the usual theory (invoking aluminum) does not altogether correspond to the observations (Figure 40), even using an almost up-to-date model like the Bilderberg model, which includes a very flat temperature minimum of 4600° (Figure 35c). To be sure, if Bonnet's observations refer to the region where the temperature is beginning to rise in the direction of the photosphere, the measured discontinuity must be interpreted as due to some other phenomenon – for it is impossible, with this model, to reproduce the discontinuity as a result of aluminum absorption.

Our knowledge of the *chromosphere* will also certainly be improved by space research. Thus, in principle, the data to be obtained from the Lyman α line and from the magnesium doublet (at 2900 Å) can be considered completely essential for establishing correct models of the chromosphere. We must, however, realize that definitive models of this type have not yet seen the light of day!

As for the X-rays (and therefore the *coronal* regions), one of the first results of space spectroscopy was the determination of new values for the abundances of the elements, using ultraviolet and X-ray lines; thus Pottasch (using permitted lines of highly ionized elements) found abundances rather different from those measured in the photosphere. His results have been confirmed by observations, in the visible region, of the forbidden lines of the corona; this important discovery has been interpreted by Delache as the result of thermal diffusion (which drives the lighter ions away from the hot regions) tempered by the solar wind. In this way one can predict quantitatively the abundances observed by Pottasch. Using these observations in the X-ray region, and assuming that the different atoms, with different ionization properties, are not formed in the same regions, Goldberg obtained a model of the transition zone between the chromosphere and the corona. We must point out that this is only a provisional model: for the inhomogeneities in these regions are certainly very great, and the calculation assumes spherical symmetry, which obviously contradicts the observations. But this model represents an important step, for it demonstrates the rapidity of the temperature increase between the chromospheric regions and the coronal regions, an effect that can take place regardless of the geometric conditions.

Concerning the *active Sun*, it is clear that energy measurements, especially in the X-ray region, enable us to determine the temperature of the active regions of the corona; this is an important point, although it has undoubtedly not yet been fully related to the other coronal observations. We can say that in order to explain the spectrum observed in active regions, it would be necessary to have plage temperatures on the order of 3×10^6 K and an electron density on the order of 10^{10} electrons per cubic centimeter. From our present position, we have the impression that there is considerable information contained in the X-ray spectrum of active regions but that – for lack of sufficient resolving power and also of a sufficiently correct intensity calibration – it is not yet realistic to attempt to take complete advantage of it.

The reader may be surprised that the author of these lines devotes so few pages to a subject as important as the ultraviolet and X-ray observations of the Sun. To be sure, this is a rapidly developing subject; but that is not the real reason for our reticence. The problem we must always keep in mind when faced with results obtained by space research, is that these results *cannot be used alone.* In a book concerned with space astronomy, it is thus important to show what purpose such and such a result may serve; but it is equally indispensable to indicate that an autonomous interpretation is impossible, and that deriving from the observations something concerning the structure of the Sun or the solar activity is a procedure involving the matching up of observations in the radio region, the space region, the visible region, and the infrared region. It is a complicated operation requiring an extremely precise knowledge of theoretical physics. The rational development of all the results of space research would be equivalent to complementing this book with another book about the Sun, for example, so great would be the number of pages to be devoted to this discussion. It is therefore important for us to emphasize what *can* be done with space observations, rather than discussing what they *have* contributed to our knowledge of a particular body – a discussion which, in our opinion, is more appropriate for a book on the Sun than for a book on space research. Let us not forget that although a given branch of science is defined *in the beginning* by particular techniques, it tends to evolve towards a synthesis which can be made only by 'object' to be studied.*

2. Galactic and Extra-Galactic Sources of Short-Wavelength Radiation

A. GAMMA-RAY SOURCES

We have seen in the previous chapter what physical processes are capable of giving rise to gamma radiation in the universe. A certain number of observations have been made by different groups in the United States, concerning gamma rays of relatively low energy – less than 1 GeV; unfortunately the observations made thus far do not appear sufficient for us to say that gamma astronomy has contributed any important results. But it remains true that in the future, information obtained from the study of gamma rays should be very important.

B. X-RAY SOURCES

On the other hand, X-ray sources are beginning to be widely known, and the information derived from them is already taking on considerable importance.

Let us first recall that the atmosphere presents a 'certain' transparency in the X-ray region, and permits (see Part I, Chapter II) observations by balloon – at present, an important opportunity (see Figure 5a). It was in 1963 that Giacconi, Rossi, and their co-workers observed the first X-ray sources beyond the Sun; and ever since 1966, studies like that of Friedman and his co-workers have been discovering a certain number of important sources (Figure 45).

It is clear that the interstellar medium itself absorbs in the X-ray region, so that the X-ray sky is certainly inaccessible at very large distances. Figure 46 shows the absorption of photons in interstellar space, due principally to hydrogen, helium, and the various metals; we see from the 'total' curve that the mean free path of a photon in the keV energy range is only on the order of 10^3 light years, and that consequently the interstellar medium is highly absorbent: the nearest stars might be visible at 80–100 eV, but the Galaxy can be crossed only by photons of higher energy than about 1.6 keV. No doubt this circumstance does offer one intriguing possibility: that of studying the absorption of the interstellar medium itself, by observations of extra-galactic sources of known properties (known from theoretical considerations, or from a reasonable theoretical extrapolation of their properties in the visible region). But at the present state of the art, interstellar absorption is more troublesome than useful, in that it prevents us from getting a clear idea of the spectra of the most distant objects.

---

* In this spirit, the author of these lines considers it necessary to abolish the national and international commissions devoted to the *results* of space research (as well as to those of radio astronomy), or to restrict their activity to the study of the techniques in question: This is the case with Commissions 40 and 44 of the IAU; it is also the case (*horresco referens!*) with COSPAR and even, in some respects, with such agencies as NASA in the United States or CNES in France. These could possibly (at least in some not too distant future) be replaced, in the scientific planning of programs, by the same institutions or committees as are charged with the planning of astrophysical and geophysical research.

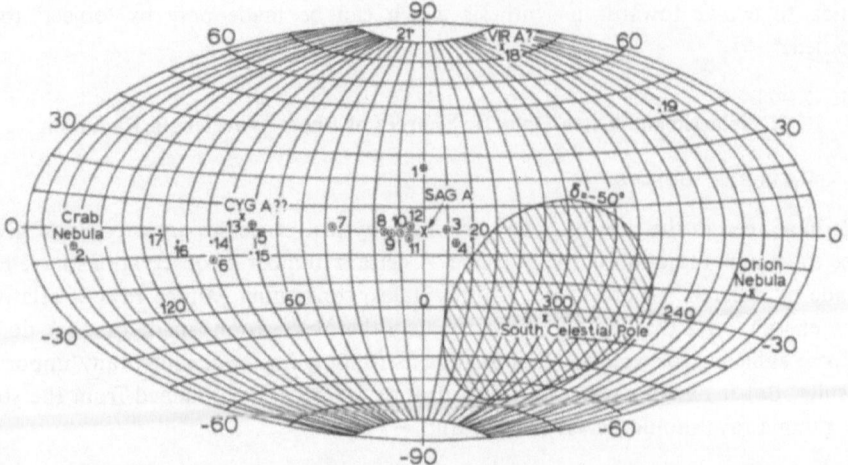

Fig. 45. The sky as observed in X-radiation (after Morrison). – Confirmed sources: ⊕; probable
sources: ⊙ ; doubtful sources: ●; interesting radio or optical objects: ×. The hatched region has
not been observed.

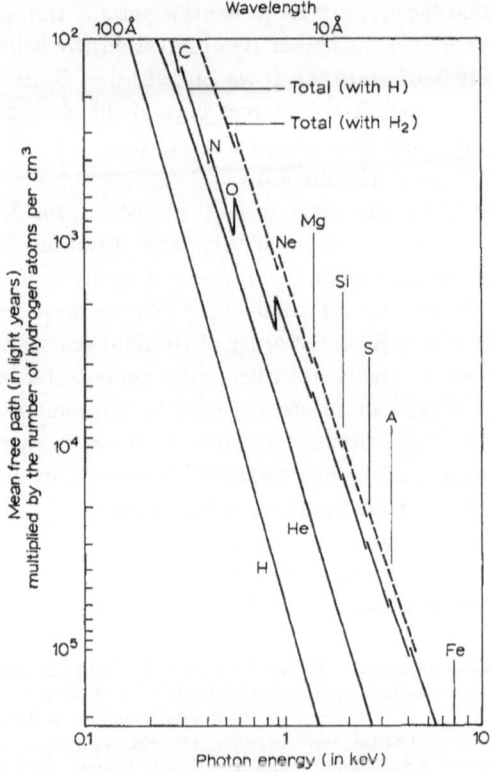

Fig. 46. Absorption of X-rays by the interstellar medium (after Morrison). The K-discontinuities
of various metallic nuclei are indicated.

Therefore the first question to be asked is: how does the sky appear at X-ray wavelengths to a terrestrial observer? We must realize that the entire sky has not yet been completely explored. However, a certain number of X-ray sources have been observed, and it is only in the vicinity of the South celestial pole that there is a notable lack of observations. The map in Figure 45 indicates our knowledge of the X-ray sky as of the end of 1966, with about twenty sources known. They are, we see, mainly concentrated near the galactic equator; of these twenty sources, only four are well enough known and sufficiently well studied for us to be sure of their existence. The intensities of these sources (with the exception, of course, of the Sun) cover a range no greater than 3 or 4 magnitudes; obviously we are dealing only with the brightest of them. Note that they are certainly very luminous objects because, at energies of several keV, the quiet Sun is fainter than the brightest X-ray source; at 1 keV, the Sun is 2 or 3 orders of magnitude brighter than the brightest extra-terrestrial source, while in the visible region the difference is 8 to 10 orders of magnitude. From these considerations and from measurements of the energy of the known X-ray sources, one can deduce that their luminosities are on the order of 10^4 times the total power of the Sun at all warelengths. It is therefore quite clear that these new sources are of considerable importance, if only because of the energy they emit. When the normal galactic radio sources were discovered, the energy in question was relatively small; but there were non-thermal emission processes, which lent importance to the study of these sources. With the X-ray sources, however, not only are we dealing with emission processes that are probably new – or at least new to the astronomical universe – but also with an emission that has a considerable effect on stellar evolution, if only because of the large flux concerned. It is thus certain that X-ray sources – much more than radio sources – are essential for the understanding of galactic evolution. For this reason, considerable efforts are currently being made to obtain a most complete and detailed study of the X-ray sources. Table XII lists the positions of the 21 sources whose existence is considered probable or possible. Several essential problems arise: first of all, *what are these sources*? Are they objects already known by other techniques, or are they new sources? Next, what is the distribution of their radiation as a function of wavelength – in other words, what is the spectrum of these sources? If they are objects that have not yet been identified, what is their *nature* and what may be their *properties*, particularly their *dimensions*?

To these various questions, we of course know only fragmentary answers; and these answers pertain only to the best-known sources.

1. *Identification*

The source Scorpius X-1 and the Crab Nebula (which is also an intense radio source) have now been correctly identified. The Crab Nebula, we know, is the remnant of a relatively recent supernova explosion. The source Scorpius X-1 is also a galactic object: it is a peculiar star. But note that certain other supernova remnants like SN 1604 in Ophiuchus and SN 1672 in Cassiopeia, which are known radio sources, are not X-ray sources at the presently detectable level; likewise, there is no X-ray

TABLE XII

List of X-ray sources in Figure 45

Number and designation	Position		Intensity keV/cm⁻² sec⁻¹	Remarks
	Right ascension	Declination		
1. Sco X-1	16^h15^m	$-15°.2$	~ 100	Identified (see text)
2. Tau X-1	$5^h31.5^m$	$+22$	~ 15	Identified with the Crab Nebula
3. Sco X-2	17^h08^m	-36.4	~ 10	
4. Sco X-3	17^h23^m	-42.2	~ 5	
5. Cyg X-1	19^h53^m	$+34.6$	5 to 15	Intensity variable
6. Cyg X-2	21^h43^m	$+38.8$	~ 5	
7. Ser X-1	18^h45^m	$+5.3$	~ 5	
8. Sgr X-2(?) (L5)	18^h03^m	-20.7	~ 5	
9. L6	18^h11^m	-17.2	~ 5	
10. L7	18^h14^m	-14.3	~ 5	
11. Sgr X-1(?)	18^h02^m	-24.9	~ 5	
12. Sgr	17^h44^m	-23.2	~ 5	
13. Cyg X-3	19^h58^m	$+40.6$	~ 5	Possible identification with the radio source Cyg A
14. Cyg X-4	21^h21^m	$+43.7$	~ 5	Possible identification with the Cygnus Loop
15. Cyg X-5	20^h40^m	$+29$	~ 5	
16. Lac X-1	22^h40^m	$+54$	~ 5	
17. Cas X-1	23^h21^m	$+58.5$	~ 5	Possible identification with the radio source Cas A
18. Vir X-1	12^h28^m	$+12.7$	~ 5	Possible identification with M87
19. Leo X-1	9^h35^m	$+8.6$	~ 5	
20. Ara X-1	16^h52^m	-46.6	~ 5	
21. Coma X-1	13^h	$+28$	~ 10	Possible identification with the Coma cluster of galaxies

source in the Cygnus Loop. The supernova remnant Cassiopeia A is a possible identification for source No. 17 on our list, but the degree of certainty is not entirely convincing. The other sources are unidentified; but their distribution near the galactic plane, with a marked concentration, suggests that they are galactic sources distributed within a few kiloparsecs of the galactic center. Their space distribution resembles that of novae or of other objects in intermediate population (between galactic Population I and Population II). In any case, it is remarkable that the physical center of the Galaxy (which we know is very close to the radio source Sgr A) is not an X-ray source to within the present accuracy of detection. Note that there are only two sources suggested at high galactic latitudes – that is, at appreciable distances from the galactic plane. For these two sources we might seek an extragalactic interpretation – in particular, for one of them, an identification with the extragalactic nebula M87 (but this is only a very provisional identification).

2. *Measured Dimensions*

As for the dimensions, it is clear that the resolving power of the instruments is rather small, and that this limits our information. The only source reported to have an appreciable size is No. 21, which has a diameter of perhaps 3°; all the other sources are smaller than a few minutes of arc. It is possible that the X-ray source located in the Crab Nebula is 1 or 2 min of arc in diameter: this has been demonstrated as a result of the occultation of this source by the Moon; but this is, obviously, an unusual and very difficult measurement. We should, however, point out that the identification of the Scorpius source (an identification which is almost certain) with a 13th magnitude, particularly blue star, implies that this source has dimensions of the same order of magnitude as those of the star itself; it is interesting to note that this star resembles an old nova, and has an intense, complex and variable spectrum of emission lines. Its visible spectrum is variable, and this variation has been measured since 1896. Thus we certainly have here a completely exceptional object.

3. *Spectrum*

Clearly, it has been possible to study only a few objects at several wavelengths. Figure 47 presents the spectrum as it is now known for the Scorpius source, the Crab Nebula, and the sum of a certain number of sources in the Cygnus region. This diagram combines the results of many observations.

How can we explain these spectra? Two theories are possible, it seems: that of synchrotron emission (the radiation of high-velocity electrons in a strong magnetic field) and that of a dilute thermal plasma. An examination of the data shows that a uniform plasma does not account for the observations, but what reason have we to suppose that the plasma is uniform? On the other hand, the energy distribution corresponding to synchrotron radiation does not agree very well with the higher energies – that is, the shortest wavelengths – measured in the Scorpius source: but the theoretical idea of a synchrotron spectrum is somewhat oversimplified, and no doubt it is no longer valid at rather large energies. We can say at present that no

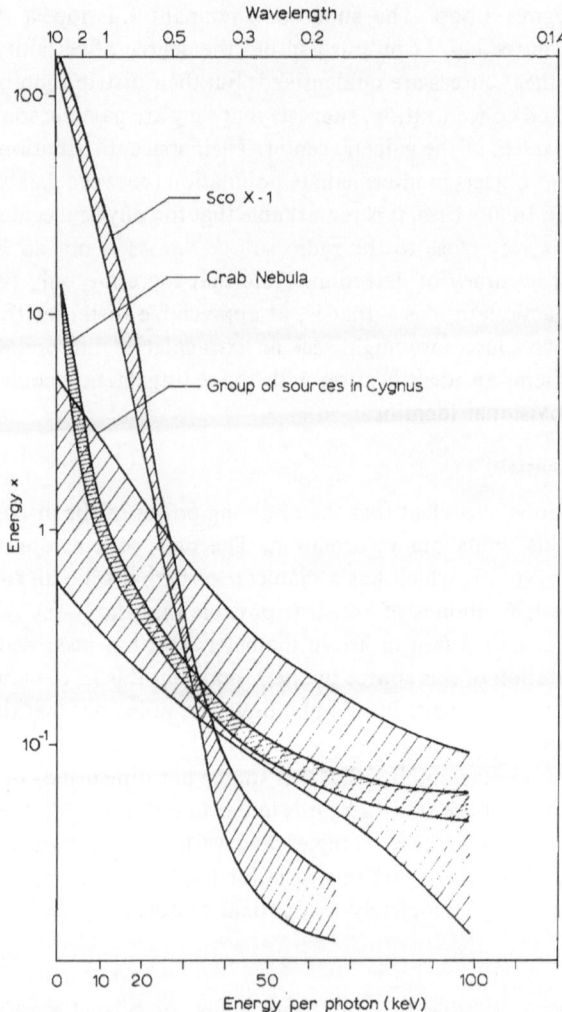

Fig. 47.   Spectra of some X-ray sources.

theory has been established to explain the spectra of X-ray sources, although we know more or less the direction that theoretical research on this subject should take.

The interpretation of Sco X-1 as a thermal source is more reasonable, considering that it is not a known radio source and that its visible spectrum more or less corresponds to the X-ray spectrum under the hypothesis of thermal radiation; perhaps we are dealing with a corona like that of the Sun.

For the Crab source, an extremely well-known and very intense radio source, the problem is no doubt different, and the interpretation by synchrotron radiation is much more probable.

C. BACKGROUND RADIATION

Besides the discrete sources we have been discussing, it is rather clear that the sky itself radiates in the X-ray region. Figure 48 shows the sky background radiation at different energies. Part of this radiation comes from stars, and part of it no doubt comes from the galactic halo; but the observations are probably principally due to radiation from the metagalaxy itself. This is only an hypothesis, for the calculations which determine the galactic radiation proper must be regarded with caution, being an extrapolation from what we know of the few sources that have been observed so far.

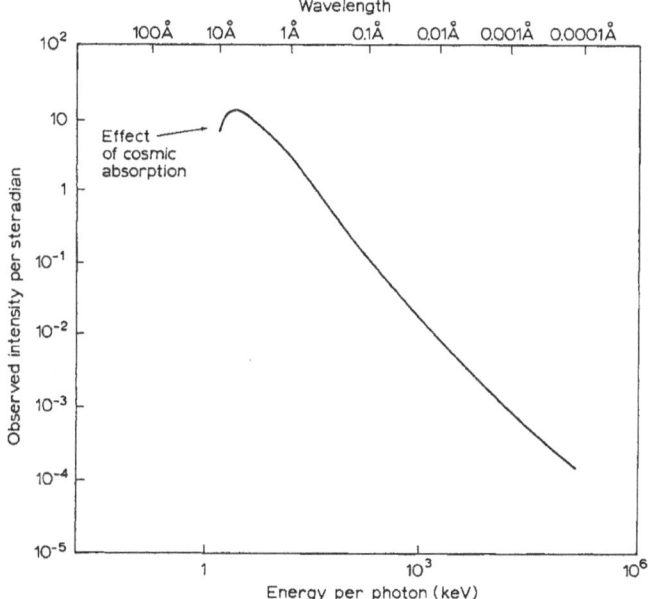

Fig. 48.   X-ray spectrum of the sky background.

D. ULTRAVIOLET SOURCES

Several types of observation have been made. First of all, one can obtain images of the ultraviolet sky; some of them show very clearly (this is the fundamental work of Kupperian and his co-workers) the existence of a certain number of high-intensity nebulous regions (Figure 49). Courtès, in a more recent study, rediscovered the same nebulosities. Unfortunately, when the experiments were repeated (under conditions which we are not absolutely sure were identical!) the nebulosities in question could not be found (either by Courtès or by Kupperian). In the present state of affairs, it is rather difficult to say whether these nebulosities have an objective reality or whether on the contrary they are caused by some still undetermined instrumental effect. The author of these lines is inclined to think that – if the origin of the nebulosities is, as he believes, to be found in the existence of a circumstellar cloud of scattering dust

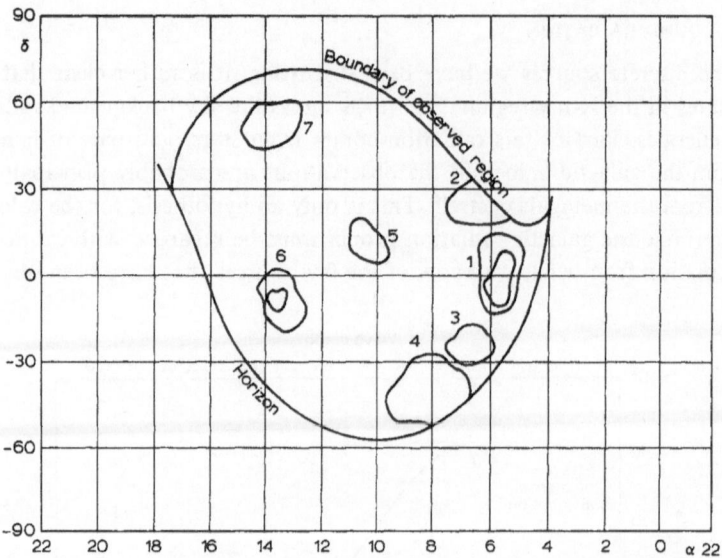

Fig. 49. Nebulosities observed in the ultraviolet, at around 1300 Å (according to Kupperian *et al*.). – Regions 1 and 6 are the brightest; 4 is also very bright; 2, 3, and 7 are of medium intensity; 5 is very weak. *Abscissa and ordinate:* equatorial coordinates in hours and degrees, respectively.

particles with very well-defined dimensions – the observations must be extremely sensitive to wavelength, and a slight change in the pass-band in which the observations are made is perhaps sufficient to conceal or to reveal the existence of the nebulosity.

Among the nebulosities discovered in this way (and, once again, of doubtful reality), we must cite those surrounding Spica (α Virginis) and the O stars of the constellation Orion. These are young, very bright stars, which certainly produce a large radiation pressure. We should also mention a certain nebulosity observed by Courtès around Sirius; on the same photograph, Jupiter is not surrounded by a nebulosity, which tends to give some credibility to Courtès' observation of a nebulosity around Sirius. Sirius is, of course, much closer to us than the Orion stars, and the nebulosity may therefore be much smaller in absolute dimensions. Finally, a third nebulosity was observed on Courtès' plate, in a region which does not seem to surround any particularly remarkable star. The characteristic of these nebulosities is their physical extent of several degrees on the sky, which implies a very considerable extent in space. Confirmation is obviously necessary before making any assertions...

A second kind of observation was made spectroscopically by the NASA team of Stecher and Milligan. These investigators showed that the spectra of hot stars were considerably depressed at wavelengths of less than 2400 Å, in comparison with the theoretical spectra derived from model atmospheres. At around 2000 Å, the reduction factor was on the order of 30. Since these first observations, the model atmospheres have been somewhat modified by the introduction of better methods for taking the absorption lines into account in the calculations. On the other hand, the measurements have been repeated and no longer give such convincing results.

These experiments dealt with continuous spectra, but the Princeton University scientists, Spitzer and Morton, have published some very remarkable spectra of hot stars, showing absorption lines in the ultraviolet (see Figure 50). In particular, the spectra of δ and π Scorpii were obtained at wavelengths greater than 1260 Å, and 29 absorption lines have been measured in these two B stars; some of these lines (Si IV, C IV) have been identified as of stellar origin, but others (O I, C II, Si IV, Al II) are probably due to interstellar absorption. In a second experiment, the same authors succeeded in photographing the spectra of six stars in Orion, at wavelengths greater

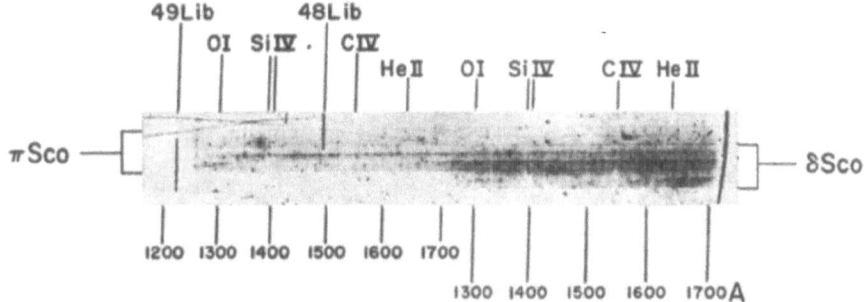

Fig. 50.   Ultraviolet spectrum of two stars: π Sco and δ Sco (after D. C. Morton and L. Spitzer, Princeton University Observatory, NASA). Note the Si IV doublet which is especially visible on both spectra.

than 1200 Å: the stars δ, ε, ζ, η, ι, and κ Orionis. In this case, emission lines were observed and identified as belonging to the ions Si IV and C IV. A remarkable peculiarity of these spectra (whose interpretation remains difficult) is the following: the intense absorption lines of Si IV and C IV in three of these spectra are strongly displaced towards the violet, implying velocities on the order of about 2000 to 4000 km/sec. These are velocities of approach, since the displacement is towards the violet; no doubt they can be attributed to gaseous shells escaping from the stars at high velocities (a sort of strong stellar wind). But it is quite clear that this important and fairly recent discovery (the launch took place in October 1965) is still in a preliminary stage of theoretical interpretation, and still poses many problems. A third flight took place in May 1966, obtaining a spectrum of the star ζ Ophiuchi. And in September 1966, new spectra were obtained of the Orion stars; these spectra confirmed the previous results. In addition to the lines observed before, a certain number of new absorption lines were also observed: for example, the lines of C III.

In another type of experiment, the magnitude measurements first made by English groups (Heddle and his co-workers) in the southern hemisphere showed, like the first measurements of Stecher and Milligan, an important absorption in the ultraviolet (Figure 51). The same type of determination of ultraviolet color indices was made more recently in the United States by means of satellites, and a color index was calculated for many stars (with the ultraviolet magnitude corresponding to a wavelength of 1427 Å). This study showed that the interstellar extinction rises to several magnitudes in the ultraviolet, on a scale where the $B-V$ extinction is equal to one unit of magnitude. It is therefore clear that there is a large interstellar absorption in the ultraviolet (Figure 52).

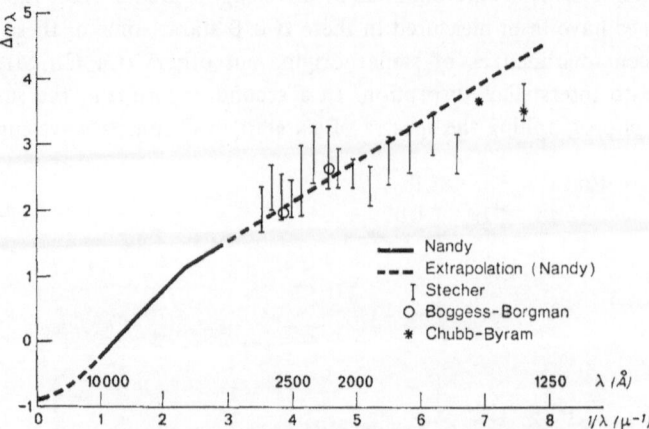

Fig. 51.   Interstellar reddening spectrum (after C. Grevesse). – The solid curve results from classical astronomical measurements. The vertical lines, circles and asterisks correspond to ultraviolet rocket measurements. *Ordinate:* the magnitude difference between an unreddened star and one which is assumed to be lowered by one magnitude in the yellow (the scale is thus arbitrary).

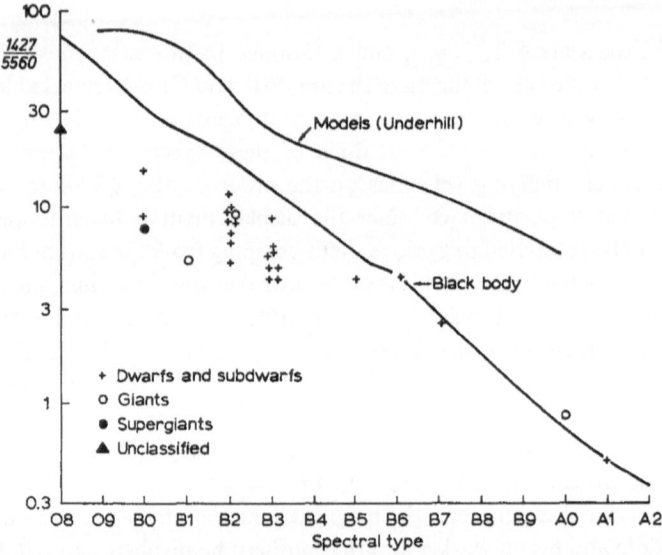

Fig. 52.   Anomalous reddening of hot stars (after C. Grevesse). – The curves represent simple (black body) and elaborate (Underhill's models) theories; the points, crosses, and circles represent the measurements. *Ordinate:* a color index between the visible ($\lambda 5560$ Å) and the ultraviolet ($\lambda 1427$ Å); *abscissa:* spectral type – hot stars at the left, cooler stars towards the right.

TABLE XIII

Intensities of the strongest emission lines in the ultraviolet spectrum

$\lambda(\text{Å})$	Identification	Erg cm^{-1} sec^{-1}
1892.03	Si III	0.10
1817.42[a]	Si II	0.45
1808.01		0.15
1670.81	Al II	0.08
1657.00[a]	C I	0.16
1640.47	He II	0.07
1561.40[a]	C I	0.09
1550.77	C IV	0.06
1548.19	C IV	0.11
1533.44	Si II	0.041
1526.70	Si II	0.038
1402.73	Si IV	0.013
1393.73	Si IV	0.030
1335.68	C II	0.050
1334.51	C II	0.050
1306.02	O I	0.025
1304.86	O I	0.020
1302.17	O I	0.013
1265.04	Si II	0.020
1260.66[a]	Si II	0.010
1242.78[a]	N v	0.003
1238.80	N v	0.004
1215.67[b]	H Lyα	5.1
1206.52	Si III	0.030
1175.70	C III	0.010
1139.89[a]	C I	0.003
1085.70[a]	N II	0.006
1037.61[a]	O VI	0.025
1031.91	O VI	0.020
1025.72	H Lyβ	0.060
991.58[a]	N III	0.010
989.79[a]	N III	0.006
977.03	C III	0.050
949.74	H Lyδ	0.010
937.80	H Lyε	0.005
835[a]	O II, III	0.010

[a] The line is perturbed by lines of other elements.
[b] The energy value for Lyα applies only to the central part of the line, in a bandwidth of 1 Å.
Note: The energies are those measured outside the Earth's atmosphere.

A more detailed study of the extinction of the interstellar matter between 2200 Å and 2600 Å has been made. The extinction clearly increases as the wavelength decreases. It seems that this phenomenon can be interpreted only by assuming that – in addition to the dust particles responsible for the reddening of starlight in the visible region – there are here and there some dust particles of much smaller dimensions that absorb in the ultraviolet. Although detailed studies have been made for the cases of graphite grains and 'dirty' ice, these investigations are not absolutely convincing; and the observations can be interpreted by means of extremely fine dust particles just as well as by particles of some new type. These same particles, localized around stars, have been invoked to explain the diffuse nebulae we were discussing above. For it is clear that in the region around a star, a dust particle is repelled with a force proportional to the square of its length while the forces of attraction are proportional to the cube of its length. The small dust particles are thus driven away, while the large dust particles are re-absorbed by the star. If the small dust particles originate in the condensation of gases ejected by the star, this is a strong argument for the circumstellar nature of clouds of these particles; but this would be true only for relatively young stars: the small dust particles that originated in much older stars should be found throughout the Galaxy, forming a second dust population in addition to the dust that absorbs in the visible region, and whose particles are relatively large in size. If this interpretation were true, we should undoubtedly have to make profound changes in our views as to the composition of the Galaxy; for it would then contain a non-negligible mass of matter in the form of small dust particles, a mass perhaps comparable to that of the rest of the Galaxy itself. Indeed, we note that in all these calculations certain stars, considered to be unreddened in the visible region, could very well *all* be reddened by interstellar matter without our having any objective method of realizing it. Laboratory studies of the condensation of dust from gas, and of the equilibrium dimensions of clouds of dust particles formed from arcs flashing between two pieces of solid material (iron or graphite, for example) tend to show that the privileged dimensions for formation of dust particles are very small, on the order of 100 Å (Lefèvre). If these results were to be confirmed in the laboratory, and if their generalization to the interstellar universe were justifiable, then there is no doubt that measurements of the reddening of starlight in the ultraviolet would provide an important argument, capable of determining the distribution of dust particles in space. Naturally, polarization measurements are necessary to complement observations of this type.

The study of these mechanisms also implies a dynamical study of the behavior of dust particles in the Galaxy and around stars; this study has been begun by C. Guillaume-Grevesse, and should complement the results of physical analysis on the one hand and the measurements made from rockets and satellites on the other hand. It is clear that in this field, laboratory astrophysics and space astrophysics must unite their efforts to obtain a better understanding of the observed phenomena. We have seen from the few examples cited that the importance of the problem is undoubtedly very great: for we all know that the mass of the Galaxy is very poorly known; mass estimates made from star counts and from the dynamics of stars near the Sun lead to very different values, the second value being much greater than the first. It is possible that the various counting methods or the dynamical studies are invalidated by large errors; but it is also possible that a sizeable proportion of the objects distributed throughout the Galaxy is inaccessible to our observations, and in particular to visual observations. This is certainly the case with neutron stars, but it is also certainly the case with dust particles of dimensions too small to have a noticeable reddening effect in the visible region – whence the interest of pursuing such observations. Note, however, that interstellar hydrogen, which is strongly absorbing in its Lyman continuum, limits the potential research domain to wavelengths greater than 912 Å.

CHAPTER V

BEYOND INFRARED OPACITY:
TOWARDS THE INFRARED BY BALLOON

Since the appearance of radio astronomy, the infrared has been somewhat neglected. We sometimes have the impression that this region, uncomfortably wedged between the visible and radio windows, can reveal only information of secondary importance.

This is undoubtedly an error, all the more serious because this fraction of the spectrum – hidden from terrestrial observers by the molecules of the lower atmosphere, principally water (H_2O) and carbon dioxide (CO_2) – is accessible by balloon; an altitude of 30 to 40 km suffices to give a considerable improvement.

The sources of radiation are, of course, principally the celestial bodies observable from the ground – above all, the Sun and the planets. For the *Sun*, the transition zone between the chromosphere and the photosphere influences the continuous spectrum at around 50 to 150 microns. At shorter wavelengths, the observations pertain mostly to the photospheric regions; at greater wavelengths, they pertain to the chromosphere, at least in the case of the quiet Sun. The properties (see Figures 31 and 32) of Planck's thermal radiation result in our obtaining a good 'mean'* temperature, since at these wavelengths and temperatures the radiation is strictly proportional to T. Moreover, relatively small regions – even if they are hot and active – have little effect on the observations. And finally, the interpretation of the observations is relatively simple, since the solar opacity is principally due to the free-free transitions of the negative hydrogen ion, of which physics now gives us a rather precise knowledge.

Like the terrestrial atmosphere, the solar spectrum is also rich in molecular lines, especially those of H_2O and CO_2; it is clear that the terrestrial H_2O and CO_2 prevent us from observing them. However, a measurement of the abundance of these molecules would be a valuable indication of physical conditions in the Sun. They would obviously give great weight to the measurements of the coldest regions in this inhomogeneous medium that we call the solar atmosphere, where the hot spots can more favorably be studied from other data like the ultraviolet.

The study of the *planets* can also be approached by the methods of infrared observation. For the moment, all that has been done is to observe the molecular spectra with which the infrared is so abundantly provided. Whether it is a question of ammonia in the atmosphere of Jupiter, or methane, or especially water vapor and carbon dioxide, the Earth's atmosphere – where the bands of the same water and carbon dioxide form an opaque screen – prevents direct observation; but space research will enable us to obtain chemical compositions, temperatures, etc. For the Moon and the

---

* In a recent study, P. Léna has derived an excellent model of the outer photosphere from infrared measurements.

satellites of other planets, which have no atmosphere, it is likely that infrared measurements will provide good determinations of the temperature.

We know that *infrared stars* (invisible at ordinary wavelengths) have been detected. It is clear that these cold objects – of stellar nature, or even pre-stellar, if not circumstellar – could be better studied from space vehicles.

Finally, we must mention the measurements of *cosmological* importance. First of all, we wish to study the distribution of radiation in the universe. In addition to stellar radiation, there exists an intense, isotropic radiation field whose temperature is 3 K, and which is supposed to represent an important constant in cosmology (remnant of the 'primeval' explosion of the universe, effect of the curvature of cosmological models...?). This is predicted by the theory, and seems to be confirmed by measurements carried out in the centimeter region. But it would be useful to know much more about it, for the cosmological constant in question (the present temperature of the 'furnace' of the universe) plays the role of a 'boundary condition' which we have to know in order to pin down the theory. Figure 53 shows that measurements – which

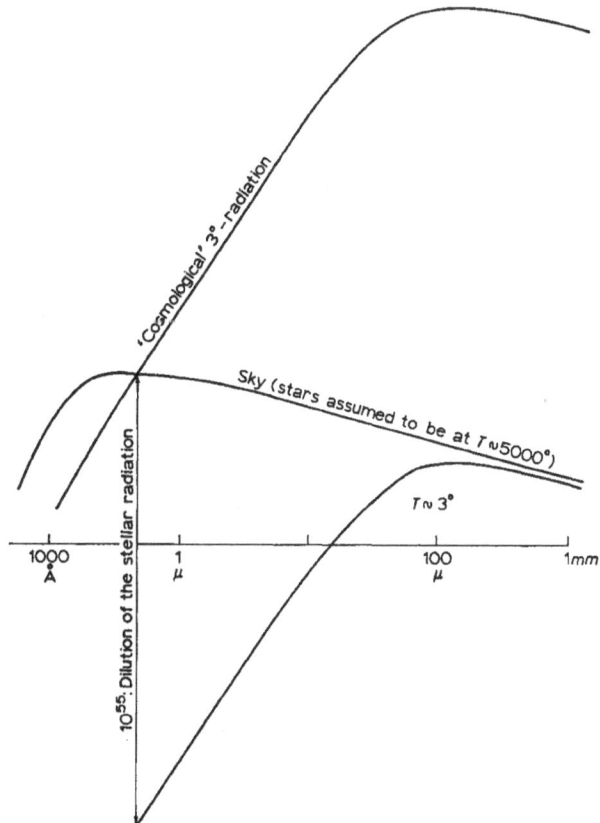

Fig. 53.   Measurement of the cosmological constant. At large wavelengths, the cosmological radiation dominates the stellar radiation.

would, however, be extremely difficult – would be most useful at wavelengths of around 100 μ or 10 μ. We have drawn in this figure the general radiation field of the Galaxy, whose color is about 10000° but whose density in the visible region is also on the order of 3°.

It is also a question of getting to know the *Galaxy* better. The abundance of hydrogen molecules H_2 is almost completely unknown. The temperature of the dust particles that clutter up the interstellar medium is also poorly known. It is also possible to detect the fine-structure components of various interstellar atoms (carbon, oxygen...).

It is undoubtedly difficult to say more. For the moment, the results are of a preliminary nature.

But the infrared, an ideal thermometer for medium temperatures (the ultraviolet gives, as we have said, the hot spots), rich in molecular bands (which are a thermometer for cold spots), is undeniably a mine of information which can be discovered only by space experiments from balloons.

CHAPTER VI

BEYOND THE IONOSPHERE:
TOWARDS VERY LONG WAVELENGTHS BY ECCENTRIC SATELLITE

We have seen that the ionosphere prevents long-wavelength radiation from the sky from arriving at the ground: this radiation is reflected by the ionized layers.

But these wavelengths, accessible only by space research, are of fundamental importance because they contain information concerning non-thermal radio sources: sources whose emission at these wavelengths would correspond to incredible temperatures, and whose origin is essentially the 'synchrotron radiation' of fast-moving electrons in magnetic fields – and perhaps some additional physical mechanisms.

We recall the relationship between frequency and wavelength:

Frequency	0.1 MHz	1 MHz	10 MHz
Wavelength	3.3 km	330 m	33 m

The most intense 'kilometer' source in the sky seems to be Jupiter. The emission of Jupiter in the usual radio astronomy region is well known: the radiation is clearly non-thermal. But a conclusion of this nature cannot be reached so lightly, even after examining Figure 54. For could there not be a localized source on Jupiter – very hot, but too small to be opaque at all frequencies, so that it is perhaps opaque only at large wavelengths? Hypotheses of this type have of course been worked out in detail. One thing is certain: the radio activity of Jupiter is linked to the solar activity – the more active the Sun, the greater the Jovian radio noise!

A detailed study which we can only summarize briefly has shown that there are three physical sources of radiation on Jupiter: first, the thermal radiation of the disk as a whole, observable only at wavelengths shorter than about 3 cm; then, the decimeter emission, which is relatively quiet but intense; and finally, the very-low-frequency bursts.

The decimeter emission could be due to a (thin) corona at $10^{5°}$; but the size of this corona would have to be much larger than what has actually been measured with interferometers. Moreover, the radiation is polarized (between 20 and 30%).

This circumstance tends to indicate that we are dealing with a directional mechanism, perhaps connected with magnetic fields localized at the emission source or in one of the media traversed by the waves. The radiation would be due to a change in energy caused by the braking (*bremsstrahlung*) of electrons rotating in a magnetic field, which emit polarized radiation in a privileged direction. The directivity and polarization of the observed radiation would obviously depend on the distribution

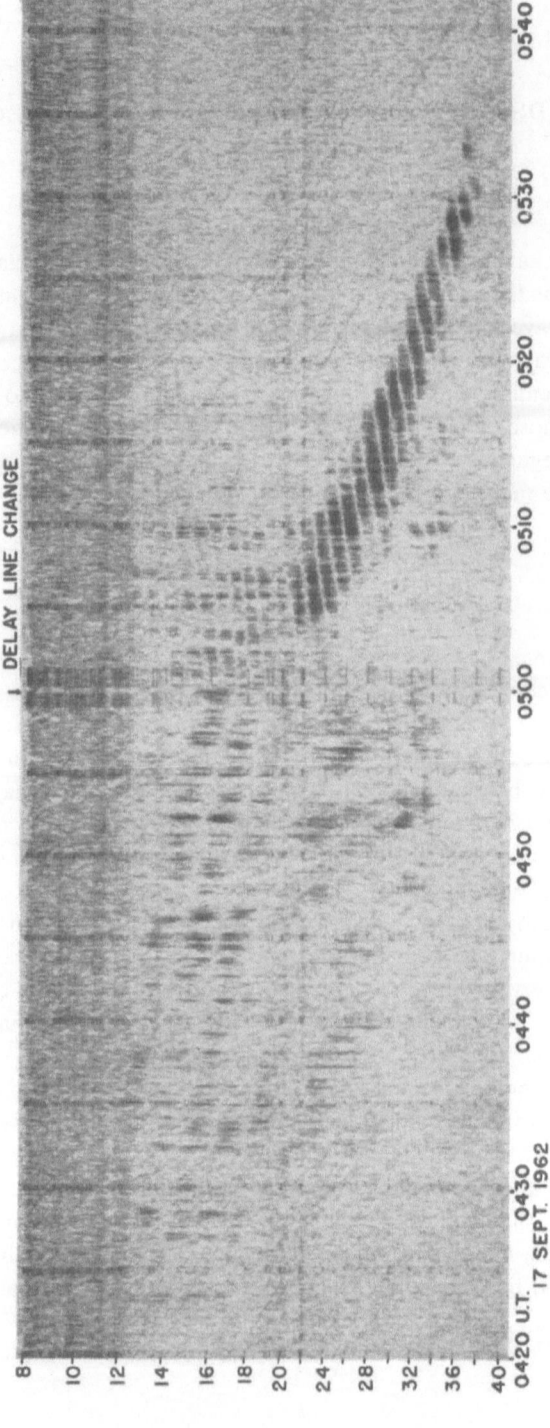

Fig. 54. Radio spectrum (of a Jovian storm) (photo by J. W. Warwick). – *Ordinate:* frequency, from 8 MHz above to 40 MHz below; *abscissa* time. These are interferometric measurements: the hatched region (the hatching is of instrumental origin) shows the progressive change in frequency of the Jovian storm.

of the magnetic field on Jupiter. In order to make such a theory agree with the deci-meter observations, it is necessary to assume a bipolar field on the order of 1200 gauss (at the poles of Jupiter), and relatively slow electrons (~ 100 keV) – a model inspired by the terrestrial Van Allen belts (see Part I, Chapter V) and due to Field. But such a model still encounters difficulties, particularly in that the predicted polarization does not agree with the observed polarization.

If one is willing to assume that the electrons have much greater energy (relativistic), than the radiation will be more strongly directional and the decimeter observations will be easier to understand. But a detailed explanation of these measurements is still problematical.

As for the very-low-frequency bursts, no exact theory is sufficiently realistic as yet; however, it is clear that localized phenomena must be assumed, and that thermal radiation can scarcely be acceptable (even though volcanic activity on Jupiter has been mentioned in this connection).

At greater wavelengths, what will happen?

Soviet satellites (Slysh, 1966) have already had an opportunity to observe Jupiter at 200 kHz. It has not been possible to locate the sources of the bursts; but in any case, the intensity emitted is in good agreement with ground-based measurements, the variability of the phenomenon decreases with decreasing frequency, and continuous radiation appears at very low frequencies.

As a curiosity, we point out that at around 200 kHz (6.6 km) Jupiter is a brighter source than the sky as a whole (almost by an order of magnitude!).

It is obvious that the study of Jupiter from space vehicles will provide a basis for theories of Jovian radiation, and therefore for a knowledge of the magnetic fields, of the Jovian ionosphere – perhaps even of the volcanic activity. It would be a good idea, of course, for the first studies (which can observe only the disk as a whole) to be complemented later on by localized research on Jupiter – perhaps working simul-taneously from the ground, perhaps at frequencies observable with some space inter-ferometer. In any case, the connection with solar activity will be interesting to follow (and easier to detect: it doubtless involves the planet as a whole!).

Note that the opacity of the Jovian ionosphere is much greater at these wavelengths than at decimeter wavelengths – as in the terrestrial case. Consequently, although the decimeter radiation may come from the lower layers of the Jovian atmosphere, this is no longer the case with very long wavelengths. This fact may explain the more continuous appearance already mentioned as characteristic of the lowest frequencies.

The other celestial radio source that is important at large wavelengths is the *Sun*, and particularly the non-thermal bursts of solar activity.

We are all familiar with the classification of these bursts, which have different physical origins. This classification is reviewed in *Plasmas et Milieux Ionisés* by E. Schatzman (Figure 10), while the interpretation of the bursts (as effects of the propa-gation through the corona of particles ejected from active regions of the Sun) is reviewed in *L'environnement de la Terre* by F. Delobeau.

At low frequencies, the 'type I' bursts are probably not observable: it seems that their intensity, like the frequency of their occurrence, decreases with decreasing frequency.

On the other hand, the type II bursts, which correspond to a rather slow frequency drift (that is, a low altitude drift of the source) should be observable at low frequencies: 30% of these bursts are observable below 25 MHz.

Type III bursts, whose drift is *very* rapid, should also be observable at low frequencies. They correspond to a displacement of the exciting agent with a velocity on the order of a third of the velocity of light! Some of them have been observed at 1 MHz.

The lowest frequencies obviously correspond to the outermost layers of the corona: the deceleration of the exciting agent in type IV bursts is one of the potential subjects for study at long wavelengths, and the nature of these bursts could then be clarified.

It is also clear that astronomical exploration of these outermost regions of the corona (20 to 40 solar radii) should be very rich in important results: for this is the transition zone between the corona proper and the solar wind – to the extent that a distinction can be made between them.

One method of observing the corona, here as in the visible region, consists of studying the occultation of a radio source by the solar corona. In the visible, it was the occultation of the Crab Nebula that made it possible to determine the characteristics of the irregularities in the corona, by interpreting the apparent size of the source as diffracted by these irregularities. In the region we are discussing at the moment, Jupiter will of course be the source to be used. Moreover, the apparent diameter of Jupiter can also be used for a study of the inhomogeneities in the interplanetary medium, whatever the position of the planet in the sky.

This research is obviously limited to the wavelengths that can pass without difficulty through the interplanetary medium. The critical frequency of the interplanetary medium is on the order of 30 kHz: this frequency represents the temporary limit of extra-terrestrial radio astronomy... until we escape from the influence of the Sun. But that is still a long way off!

The purpose of solar radio astronomy at frequencies between 10 MHz and 30 kHz will therefore be to study the spectrum of bursts and to locate them (it is possible to use interferometry, taking advantage of reflection from the upper layers of the ionophere), and thus to study the outer layers of coronal plasma, which are relatively close to the Earth.

Yet a third type of radio-astronomical research at low frequencies is possible: this is the study of the radiation coming from the *Galaxy* itself, or from *extra-galactic radio sources*. Ground-based measurements completely exclude the possibility of thermal radiation at the frequencies in question.

As far as the Galaxy proper is concerned, this radiation is also produced by the braking of very high-speed electrons in a magnetic field (what we call – have I already mentioned it? – the 'synchrotron' effect); the theory of this phenomenon requires a flux density proportional to $v^{-0.6}$, which more or less agrees with the observations between 10 and 1000 MHz. There is a certain arbitrary factor in the theory: the energy distribution of the electrons responsible for the radiation. Now, if this parameter is fixed in such a way as to 'force' agreement between theory and experiment, the resulting energy distribution of the relativistic electrons is similar to that obtained for cosmic radiation: this we can be fairly confident of the theory.

Towards low frequencies, measurements are almost non-existent. But they have made it possible to discern a band of ionized hydrogen in the galactic plane: we are therefore dealing with emission of a different kind. But the intensity measured is proportional (at a given wavelength) to the product of the square of the number of electrons by the length of the absorbing path, for the total opacity is small. It is clear, in spite of the paucity of the measurements, that the $v^{-0.6}$ spectrum is perturbed: the observed spectrum is almost flat. A decrease in brightness with frequency seems to have been observed: is absorption by hydrogen between the source and the observer responsible? What is the electron density in the media responsible for the absorption? What is the emission mechanism? Where does it take place? All these questions present themselves to the space radio astronomers who plan to work between 30 kHz and 2 Hz.

Space radio astronomy is certainly still in swaddling clothes. The study of other planets besides Jupiter, of the interplanetary medium, of the solar corona and the solar wind, of our Galaxy (and perhaps of distant extra-galactic sources) – what a splendid program! But not just any satellite can be used. This is above all a task for space probes which go beyond the ionized layers of the atmosphere; or, better still,

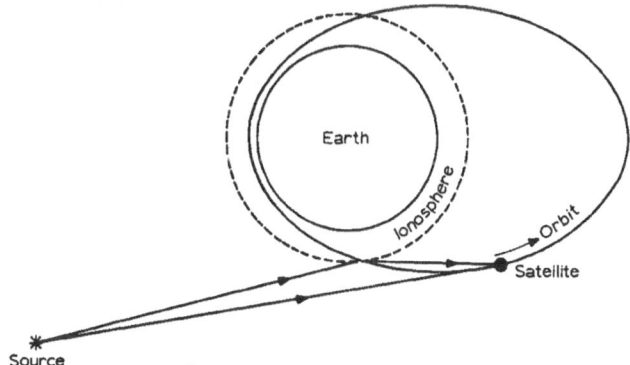

Fig. 55. Possible use of a space probe. – A source is studied by means of the interference pattern obtained when its radiation is reflected from the ionosphere.

for very eccentric satellites (whose apogee is several tens of thousands of kilometers, if not 100000 or 200000 km, above the surface of the Earth) which can take advantage of their passage through the ionosphere to separate out certain extraneous effects, or to make interplanetary measurements by using the interference patterns produced by the reflection of kilometer waves from the ionosphere (Figure 55). The reader can imagine the difficulty of such enterprises.

CHAPTER VII

A PROSPECTIVE CONCLUSION OF LIMITED PERSPECTIVE

I announced at the beginning of Part II of this book, that I had no intention of producing a complete inventory (for lack of space, if not of imagination) of everything that could be done if there were no atmosphere, or if we could escape it with impunity... Eliminate diffusion, and reach for faint objects – eliminate absorption, and make use of gamma rays, X-rays, the ultraviolet, the infrared, and kilometer radio waves. We have devoted most of Part II to an exploration of these possibilities.

Eliminate the terrestrial magnetic field, and study the local environment – we have said a few words about this in Part I, Chapter V.

Eliminate scintillation – and then the potential of gigantic instruments will enable us better to separate close objects: double stars of small separation, or at great distances; planets, and details of planetary surfaces. This is still virgin territory, and will no doubt remain so for some time (see Part I, Chapter IV)...

But let us enlarge our perspective. The Earth is not the only obstacle. The solar environment of dust and gas, and the interplanetary field, hide from us an important part – perhaps an essential part – of the galactic radiation...

The interstellar medium, the galactic halo, and the interstellar dust hide from us an important part – perhaps an essential part – of the metagalactic world...

And does not the metagalaxy itself – where Zwicky, in discovering bridges between galaxies, has proved that scarcely a vacuum exists; where clusters of galaxies are stuffed with diffuse matter; and where magnetic fields, although very weak, are effective because of their enormous extent – hide from us some essential aspects of the cosmology of our universe?

Now, we must realize that the successive steps will be difficult, as these restrictions are not of the same order of magnitude. For the moment, and no doubt for a long time to come, it is the terrestrial atmosphere from which we are trying to escape. In other words, in order to know the rest – that which remains hidden even after the conquest of interplanetary space – we must still call upon physics, upon theory. Phenomena observed from the surface of the Earth or from the surface of the Moon will only be signs and indications of more important phenomena, and there too theory will have to make the necessary inductions. In the conquest that has barely begun – and this is the point of the present conclusion – it would be erroneous to believe that rockets are our only tools. Like the blind man and the paralytic of the fable, the techniques of theory and those of space experimentation remain inseparable. May the reader excuse me for insisting once more upon this elementary verity!

Note. Just as I finish correcting the proofs of this book, the Apollo astronauts have returned from their landing on the Moon and the latest Mariner is taking pictures of Mars. Although these exploits (sometimes more sporting than astronomical) open the way to *exploration* of the nearest world, and of our neighborhood in space – see *Experimental Astronomy* by the same author – they also presage the great *observatories* which, without a large decrease in their distance from the objects to be studied, will escape from the terrestrial atmosphere. I hope that the continuous progress in our knowledge of the universe will be comparable with the efforts expended in this 'big show', and the funds and energy that have been invested in it.

CONCLUSION

Space astronomy is still in the exploratory stage. The data obtained – whether pertaining to the Sun or to distant galaxies – often provide the analyst only with orders of magnitude. But this has already been enough to modify a number of opinions concerning these well-known objects. On the other hand, extension of the observations to very short wavelengths has enabled us to discover and study new objects, previously unknown. What riches may we not expect from the gain in pointing accuracy and refined photometry that space astronomy will bring, in comparison with what has already been done with ground-based telescopes? It is thus undeniable that a modern astronomer no longer has the right to ignore the potential of extraterrestrial laboratories, any more than a space astronomer can disregard the data accumulated at ground-based observatories.* New techniques can and do help us, not only to solve old problems and to refine known solutions, but also to state the old problems in a new way, and to formulate new problems. When astronomy became *experimental*, it gave us the pride of a poor man's demiurge. But in becoming *extra-terrestrial*, it teaches us on the contrary a little more detachment; it teaches us to reject one more anthropomorphism, that of the sky... up above the roof.

---

* The reader is referred to the conclusion of *Experimental Astronomy*, where the author demonstrates in more detail that the new techniques complement, but cannot replace, the techniques of classical astronomy.

APPENDIX

TABLE I
Universal physical constants

Velocity of light	$c = 2.99791 \times 10^{10}$ cm sec^{-1}
Gravitation constant	$G = 6.668 \times 10^{-8}$ dyne cm^2 g^{-1}
Planck's constant	$2\pi\hbar = h = 6.6237 \times 10^{-27}$ erg sec
Electron charge	$e = 4.80217 \times 10^{-10}$ e.s.u.
Electron mass	$m_e = 9.1071 \times 10^{-28}$ g
Boltzmann's constant	$k = 1.38024 \times 10^{-16}$ erg/deg
Avogadro's number (physical)	$\mathcal{N} = 6.0238 \times 10^{23}$
Gas constant	$\mathcal{R} = 8.3143 \times 10^7$ erg/deg
Mass of the hydrogen atom	$M_1 = 1.6734 \times 10^{-24}$ g
Stefan's constant	$\sigma = \dfrac{8\pi^5 k^4 c}{15 c^3 h^3 4} = 5.6698 \times 10^{-5}$ erg cm^{-2} deg^{-4} sec^{-1}

Electron volt (1 eV):

associated wavelength	12396.3×10^{-8} cm
associated wavenumber	8067.1 cm^{-1}
associated frequency	2.41838×10^{14} sec^{-1}
associated energy	1.60184×10^{-12} erg
associated temperature	11605.9 K

Parsec	3.0857×10^{18} cm $= 206265$ AU $= 3.262$ light years
Astronomical unit (mean Earth–Sun distance)	AU $= 1.496 \times 10^{13}$ cm $= 499.01$ light seconds
Light year	l.y. $= 9.4605 \times 10^{17}$ cm $= 6.324 \times 10^4$ AU

TABLE II
The Sun ☉

Diameter: $2R = 1.3920 \times 10^{11}$ cm
Mass: $M = 1.989 \times 10^{33}$ g

Axial rotation (sidereal):

$\lambda = 0°$: $14°.5$ per day (equator)
$\lambda = 45°$: $13°.2$
$\lambda = 90°$: $11°.8$ (pole)

TABLE III

The solar system

	Planet	Satellites (non-exhaustive list)	Diameter (D) (km)	Mass (M) (grams)	Semi-major axis of orbit, a (10⁶ km)*	Period of revolution (P) (years or days)*	Period of axial rotation (P) (days, hours)	Remarks
☿	Mercury	(None)	4840	3.333×10^{26}	57.9	0.240 y	88 d 0	No atmosphere
♀	Venus	(None)	12228	4.870×10^{27}	108.2	0.615 y	?	Thick, cloudy atmosphere
♄, ⊕	Earth	–	12742	5.976×10^{27}	149.6	1.000 y	23h 56m 4s 099	Atmosphere
		Moon	3476	7.35×10^{25}	0.3844	27.321661 d	–	No atmosphere
♂	Mars	–	6770	6.443×10^{26}	227.9	1.881 y	24h 37m 22s 668	Thin atmosphere (CO_2 abundant)
		Deimos	(15)	?	0.02352	1.262 d	–	
		Phobos	(10)	?	0.00937	0.3189 d	–	
♃	Jupiter	12 satellites among them:	140720	1.8993×10^{30}	778	11.862 y	9h 50m to 9h 56m	Thick atmosphere, polyatomic molecules: NH_3
		Io	3550	7.2×10^{25}	0.422	1.769 d	–	
		Europa	3100	4.7×10^{25}	0.671	3.551 d	–	
		Ganymede	5600	15.5×10^{25}	1.070	7.155 d	–	
		Callisto	5050	9.7×10^{25}	1.880	16.689 d	–	
♄	Saturn	10 satellites among them:	116820	5.684×10^{29}	1427	29.456 y	10h 14m to 10h 40m	Thick atmosphere; Saturn has a flat, structured equatorial ring
		Tethys	1000	6.5×10^{23}	0.295	1.888	–	
		Dione	–	1.0×10^{24}	0.377	2.737	–	
		Titan	4950	1.4×10^{26}	1.222	15.95 d	–	
♅	Uranus	5 satellites among them:	47100	8.676×10^{28}	2870	84.015 y	10h 8	
		Ariel	–	–	0.192	2.520 d	–	
		Umbriel	–	–	0.267	4.144 d	–	
♆	Neptune	(2 satellites)	44600	1.029×10^{29}	4496	164.788 y	15h 8	
♇	Pluto	(None)	6000	5.53×10^{27}	(5900)	247.7 y	6d 39	Surface probably reflecting

* Around the Sun for planets, around the planet for satellites.

BIBLIOGRAPHY

There are few text-books devoted entirely to space research. However, a large number of *symposia* have been dedicated to this subject. There follows a non-exhaustive list.

Discussion on Observations of the Russian Artificial Earth-Satellites and their Analysis. Royal Society Symposium, London, November 29, 1957. *Proc. Roy. Soc.*, Ser. A **248**, No. 1252, 1958.

Dynamique des satellites. IUTAM Symposium, Paris, May 28–30, 1962, (ed. by M. Roy), Springer, Berlin-Göttingen-Heidelberg, 1963.

Use of Artificial Satellites for Geodesy, 1st International Symposium, Washington, April 26, 1962 (ed. by G. Veis), North-Holland Publ. Co., Amsterdam, 1963.

Astronomical Observations from Space Vehicles. IAU Symposium No. 23, Liège, August 1964 (ed. by J.-L. Steinberg), Publications du CNRS, Paris, 1965; *Ann. Astrophys.*, **27** and **28**, 1964 and 1965.

Trajectories of Artificial Celestial Bodies as Determined from Observations. COSPAR-IAU-IUTAM Symposium, Paris, April 20–23, 1965 (ed. by J. Kovalevsky), Springer, Berlin-Heidelberg-New York, 1966.

International Space Science Symposia, organized by COSPAR and published under the title *Space Research*, by North-Holland Publ. Co., Amsterdam.

    I. Nice, 1960 (ed. by H. Kallman), 1960.

    II. Florence, 1961 (ed. by H. C. Van de Hulst, C. de Jager, and A. F. Moore), 1961.

    III. Washington, 1962 (ed. by W. Priester), 1963.

    IV. Warsaw, 1963 (ed. by P. Muller), 1964.

    V. Florence, 1964 (ed. by D. G. King-Hele, P. Muller, and G. Righini), 1965.

    VI. Mar del Plata, 1965 (ed. by R. L. Smith-Rose), 1966.

    VII. Vienna, 1966 (ed. by R. L. Smith-Rose, in collaboration with S. A. Bowhill and J. W. King), 2 vols, 1967.

    VIII. London, 1967 (ed. by A. P. Mitra, L. G. Jacchia, and W. S. Newman), 1968.

    IX. Tokyo, 1968 (ed. by K. S. W. Champion, P. A. Smith, and R. L. Smith-Rose), 1969.

The following *introductory books* are devoted to space research:

R. E. Jastrow, *The Exploration of Space*, Macmillan, New York, 1960.

A. Danjon and P. Muller, 'Artificial Satellites and Space Vehicles', Book VIII of *The Flammarion Book of Astronomy* (transl. by A. and B. Pagel), George Allen and Unwin, London, 1964.

A. I. Berman, *The Physical Principles of Astronautics*, Wiley, New York, 1961.

W. E. Liller (ed.), *Space Astrophysics*, McGraw-Hill, New York, 1961.

D. King-Hele, *Observing Earth-Satellites.* Macmillan, London-Melbourne-Toronto, 1966.

Z. Kopal, *Telescopes in Space*, Faber & Faber, London, 1968.

The problems of *general astronomy and astrophysics* are treated in a large number of books, at all levels and in several languages. In English, and in order (approximately) of increasing difficulty, the following might be an appropriate selection:

J.-C. Pecker, *The Sky* (transl. by L. Vogel), Orion Press, New York, 1962.

G. Abell, *Exploration of the Universe*, 2nd ed., Holt, Rinehart and Winston, New York, 1969.

A. Unsöld, *The New Cosmos* (transl. by W. H. McCrea), Springer-Verlag, New York, 1969.

T. L. Swihart, *Astrophysics and Stellar Astronomy*, Wiley, New York, 1969.

J. Kovalevsky, *Introduction to Celestial Mechanics* (translated from the French), Reidel, Dordrecht, 1967.

H. Y. Chiu, *Stellar Physics*, Blaisdell, Waltham, Mass., 1968.

V. A. Ambartsumyan, *Theoretical Astrophysics* (transl. by J. B. Sykes), Pergamon Press, New York, 1958.
D. S. Evans, *Observations in Modern Astronomy*, The English Universities Press, London, 1968.
J. P. Cox and R. T. Giuli, *Principles of Stellar Structure*, Gordon and Breach, New York, 1968.
J. T. Jefferies, *Spectral Line Formation*, Blaisdell, Waltham, Mass., 1968.

On the particular problems of the *terrestrial atmosphere* and the *Sun*, the following works contain the essential bibliography:

H. Zirin, *The Solar Atmosphere*, Blaisdell, Waltham, Mass., 1967.
E. Tandberg-Hanssen, *Solar Activity*, Blaisdell, Waltham, Mass., 1967.
C. de Jager (ed.), *The Solar Spectrum*, D. Reidel, Dordrecht, 1965.
The Solar System:
 I. *The Sun* (ed. by G. P. Kuiper), Univ. of Chicago Press, 1953.
 II. *The Earth as a Planet* (ed. by G. P. Kuiper), Univ. of Chicago Press, 1954.
Handbuch der Physik, vol. LII: *Astrophysik;* III: *The Solar System*, Springer, Berlin-Göttingen-Heidelberg, 1959.

As for *galaxies* and the *Galaxy*, we recommend the following works:

Handbuch der Physik, vol. LIII: *Astrophysik*; IV: *Stellar Systems*, Springer, Berlin-Göttingen-Heidelberg, 1959.

Stars and Stellar Systems (ed. by G. P. Kuiper and B. M. Middlehurst), Univ. of Chicago Press:
 V. *Galactic Structure* (ed. by A. Blaauw and M. Schmidt), 1965.
 VI. *Stellar Atmospheres* (ed. by J. L. Greenstein), 1960.
 VII. *Nebulae and Interstellar Matter* (ed. by B. M. Middlehurst and L. H. Aller), 1968.
 VIII. *Stellar Structure* (ed. by L. H. Aller and D. B. McLaughlin), 1965.

Finally, there are many flourishing *journals* and *series* dedicated to *space research*. The fundamental articles can be found in the following journals:

Advances in Astronautical Sciences, New York.
Annual Review of Astronomy and Astrophysics, Academic Press, New York-London.
Astronautica Acta, International Astronautical Federation, Vienna, Springer.
Planetary and Space Science, Pergamon Press, London-New York.
NASA Publications (non-periodical), Washington, D.C.
Revue française d'Astronautique, Paris.
La Recherche spatiale, Centre National d'Études Spatiales, Paris.
Solar Physics, D. Reidel, Dordrecht.
Space Science Reviews, D. Reidel, Dordrecht.

This bibliography can be complemented by the excellent sources of astronomical and astrophysical *data:*

Landolt-Bornstein, Gruppe VI, *Astronomy, Astrophysics, and Space Research*, vol. I: *Astronomy and Astrophysics*, Springer, Berlin, 1965.
C. W. Allen, *Astrophysical Quantities*, 2nd ed., Univ. of London, The Athlone Press, 1963.

and by the regular *reports* of the commissions of the International Astronomical Union, especially the following commissions: Commission 9 (Astronomical Instruments), Commission 10 (Solar Activity), Commission 12 (Radiation and Structure of the Solar Atmosphere), Commission 16 (Physical Study of Planets and Satellites), Commission 17 (The Moon), Commission 28 (Galaxies), Commission 29 (Stellar Spectra), Commission 36 (Theory of Stellar Atmospheres), Commission 40 (Radio Astronomy), and Commission 44 (Astronomical Observations from Outside the Terrestrial Atmosphere).

ASTROPHYSICS AND SPACE SCIENCE LIBRARY

Edited by

J. E. Blamont, R. L. F. Boyd, L. Goldberg, C. de Jager, Z. Kopal, G. H. Ludwig, R. Lüst,
B. M. McCormac, H. E. Newell, L. I. Sedov, Z. Švestka, and W. de Graaff

p.t.o.

16. S. Fred Singer (ed.), *Manned Laboratories in Space. Second International Orbital Laboratory Symposium.* 1969, XIII + 133 pp.

17. B. M. McCormac (ed.), *Particles and Fields in the Magnetosphere. Symposium Organized by the Summer Advanced Study Institute, held at the University of California, Santa Barbara, Calif., August 4–15, 1969.* 1970, XI + 450 pp.

18. Jean-Claude Pecker, *Experimental Astronomy.* 1970, X + 105 pp.

19. V. Manno and D. E. Page (eds.), *Intercorrelated Satellite Observations related to Solar Events. Proceedings of the Third ESLAB/ESRIN Symposium held in Noordwijk, The Netherlands, September 16–19, 1969.* 1970, XVI + 627 pp.

20. L. Mansinha, D. E. Smylie and A. E. Beck, *Earthquake Displacement Fields and the Rotation of the Earth. A NATO Advanced Study Institute. Konference Organized by the Department of University of Western Ontario, London, Canada, 22 June – 28 June 1969.* 1970, XI + 308 pp.

SOLE DISTRIBUTORS FOR U.S.A. AND CANADA:

SPRINGER-VERLAG NEW YORK, INC., 175 Fifth Ave., New York, N.Y. 10011